THE HBD COOKBOOK

Life-changing recipes for long-term health and perfect weight

PETRONELLA RAVENSHEAR

Thorsons

Thorsons
An imprint of
HarperCollins*Publishers*
1 London Bridge Street
London SE1 9GF

www.harpercollins.co.uk

HarperCollins*Publishers*
Macken House, 39/40 Mayor Street Upper
Dublin 1, D01 C9W8, Ireland

First published by Thorsons 2023

10 9 8 7 6 5 4 3 2 1

A catalogue record of this book is available from the British Library

ISBN 978-0-00-860078-5

Food Stylist: Lizzie Kamenetzky
Prop Stylist: Lydia McPherson

Printed and bound by GPS Group, Slovenia

WHEN USING KITCHEN APPLIANCES PLEASE ALWAYS FOLLOW THE MANUFACTURER'S INSTRUCTIONS

THE HBD COOKBOOK

Dedication

This book is dedicated to my beloved Riccardo and to his restaurant, Riccardo's, which he opened in London's Chelsea in May 1995. Riccardo was one of the first restaurateurs in the world to offer gluten-free foods, as well as a vegan menu, and he serves the finest produce available (and in true Mediterranean style only uses extra virgin olive oil in the kitchen). Those were the things that drew me to Riccardo's many years ago, and also happily drew us together!

Contents

Introduction

This is *The HBD Cookbook*, the companion to *The Human Being Diet* (HBD), which is going to help you become the best, healthiest, lightest, most energetic, happiest version of YOU.

Before we get started, though, I'd like to share this astonishing fact from the World Health Organization (WHO). For the first time in our human history, we are dying of chronic diseases – but these are lifestyle diseases, rather than the ones we used to catch from bacteria or viruses. And we're dying of these 'non-communicable diseases' over and above all the other life-threatening dangers that we've ever faced in our ancient past – including starvation, dehydration, injuries and extremes of temperature.

These chronic diseases share two characteristics:

1) If not held in check they ultimately lead to an untimely death.

2) They all involve chronic low-grade inflammation (more about that later).

Our modern lifestyle diseases include metabolic syndrome (high blood pressure, overweight/obesity, high blood sugar levels and high levels of blood fats), cancer and chronic respiratory disease. And they kill 41 million people a year, which equates to 71 per cent of deaths worldwide. The WHO refers to the causes of these diseases as '*Unhealthy diets and a lack of physical activity*' and further states, also on their website, '*Nutrition is a critical part of health and development. Better nutrition is related to improved infant, child and maternal health, stronger immune systems, safer pregnancy and childbirth, lower risk of non-communicable diseases (such as diabetes and cardiovascular disease), and longevity.*'

We are in a worse state, health-wise, than we were in the years following the Second World War, when we faced rationing and all the other deprivations. So what's gone wrong; why are these chronic diseases escalating and what can we do to stem the tide? And what exactly does the WHO mean by 'better nutrition'? If we knew exactly what that meant, and if we also knew what was meant by the general advice bandied about to eat 'a healthy balanced diet', who wouldn't jump at the chance of feeling better and healthier and living longer?

It doesn't help that all but the most enlightened of medics are still telling us that what we eat (bar sugar, salt and saturated fat) doesn't broadly affect our health. And the general advice for weight loss – to eat less and exercise more – is particularly unhelpful. Also unhelpful is the regular barrage of contradictory diet and weight-loss headlines in the press – whether it be the latest superfood or miracle diet; although we know deep down that the magic bullet doesn't exist, we're still tempted to try to track it down. We've been lured in by countless slimming promises involving raspberry ketones, acai berries and fat blockers, and endless fad diets that just don't work in the long term.

Defining what's meant by 'better nutrition' or 'a healthy diet' in reference to the WHO statements above, is controversial to say the least. But, generally speaking, the current consensus is that the Mediterranean Diet (MD), combined with some intermittent fasting and light exercise (walking) is the way to go. The MD focus is on vegetables, fruit, legumes, nuts, seeds, olive oil and whole grains, with moderate amounts of fish and seafood, dairy, poultry and eggs, and very little red meat.

So the question is, could the common health problems that we face, such as obesity, skin disorders, pain, depression, insomnia, hormonal imbalances, digestive disorders, frequent infections, infertility, and many others, with regards to the WHO statement above, be changed in any way at all by our diet? **And the answer is yes.** And this is exactly what HBD is all about, and why many HBDers (HBD fans and followers who have read the book and followed it to the letter) have suggested that the programme be taken up by the NHS because the health benefits have been so startling.

We can change the way we look and feel by changing what and how we eat; what we choose to put on the end of our forks and into our mouths has a direct effect on our overall health and energy. The purpose of the HBD book and this HBD cookbook is to give you the tools to enable you to proactively take control of many vital aspects of your health, and to explain that we alone are ultimately responsible for our health. It's no good just going to the doctor and picking up a prescription to manage our symptoms and expecting to be magically 'fixed'. We have the power to take our health into our own hands and we can optimise it, and safeguard our future health, by using these recipes and by reading my original HBD book. These two books are your tools for success.

Introducing The Human Being Diet, aka HBD

HBD is as close as it gets to a one-size-fits-all style of healthy eating for anyone and everyone. And that's why it's called The Human Being Diet. By the way, the 'diet' in the book's title refers to the word's original meaning, which is 'way of life', rather than to a weight-loss diet.

HBD is a healthy and sustainable way of eating that is adaptable enough to suit everyone (with the exception of pregnant or nursing mothers or growing children, who need more specialised nutrition) whether they want to lose weight or not. Perhaps the most important point to make about HBD is that although many people (about 80 per cent) are drawn to the programme as a healthy means of losing weight, the weight loss occurs as a side-effect of returning the body to balance. It's as if our body heaves a sigh of relief and gradually reduces the inflammation we've inadvertently created. As the inflammation diminishes our weight diminishes, too. And it is this that sets HBD apart from other healthy eating programmes.

There's no calorie or carb counting with HBD. There are no meal replacements or ultra-processed foods. All the ingredients for the delicious recipes in this book can be found in the local supermarket – there are no gimmicks, no bars or shakes, just real, wholesome, nutrient-packed food. The kind of food that we humans evolved on, the kind of food that provides the essential nutrients we need for energy, for health, for immune strength and resilience and for life itself. A diet of crisps, biscuits and chicken nuggets will keep us alive, but is just being alive good enough? Wouldn't you also like to feel vibrantly well?

HBD is not vegan or paleo or keto or high-fibre or low-calorie or low-fat (although the first 16 days of the programme are low both in calories and in fat). HBD is a three-month programme involving four phases and 10 rules that seamlessly transitions into a new way of life. By the end of the three months, when you move into the fourth phase, you'll find you feel so well that you'll never want to revert to your old ways of eating. You'll understand what works for you personally. You will have the flexibility to make this programme all YOURS by learning to listen more closely to, and to understand the language of your body. There are two clear reasons why three months is the critical period for the plan; many experts agree that this is the minimum amount of time to consolidate new habits, and also the life of a red blood cell is around 100 days, which means this is how long it takes for your cells to renew themselves.

We're going to be eating three meals a day of wholesome, minimally processed food – like we did in the old days. HBD meals combine equal weights of protein foods (fish, meat, poultry, etc.) with vegetables. There's a five-hour fast between each meal with nothing but water consumed between meals. It sounds simple and it *is* simple! Eating the HBD way is all about giving our bodies exactly what they need in order to call a halt to the creep of chronic disease; and about maximising our immunity, energy and longevity. It can be used both for weight loss and/or for its wider health benefits.

Eating the HBD way is all about giving our bodies exactly what they need

So now we know that weight loss occurs on HBD as a side-effect of improved overall health, would this be a healthy and suitable way to eat for someone who didn't want or need to lose weight? Absolutely! Some of the conditions and symptoms that have improved with the health-giving meals in the programme, apart from metabolic syndrome, or the beginnings of Type 2 diabetes, include digestive problems such as IBS and low energy, insomnia, PMT/PMS and PCOS, infertility, skin problems, headaches and joint pain; almost any niggling problem can be eaten away with a change of diet. HBD can be followed by anyone who wants to proactively improve their health long term.

I wrote HBD with omnivores in mind because that's what the original humans were. We weren't choosy about what we ate – we ate anything, plant or animal, that we could find. But growing numbers of people who are plant-based have taken to HBD like ducks to water and are getting great results with it. Personally, I love fish and eggs and butter and cheese and (having dabbled with short periods of veganism) know that an omnivorous diet suits my body best; HBD can be adapted to suit everyone apart from fruitarians (sorry, fruitarians).

Following the HBD programme helps us to demystify our health and encourages us to discover for ourselves how the food that we eat affects how we feel. In following HBD we discover which foods best suit us, and which are best avoided, and that enables us to create our very own personalised eating programme for life. Everyone's encouraged to cut out dairy and nightshade vegetables (tomatoes, aubergine, peppers and potatoes) for the first 16 days because so many of us have problems with these foods. It's not until we completely eliminate a food for at least 16 days before reintroducing it that we notice the difference. In my experience, the body can be stubborn, and sometimes it really can dig its heels in, and it can take up to 13 or 14 days for change to take place for some people, so the 16 days gives enough time for everyone to really make the switch. Most people react more quickly than this, however.

I'm not making any health claims for HBD. I'm talking about the curative properties of basing our meals on real, nutrient-dense food. It's a hackneyed old phrase, that one from Hippocrates, but it's still true: *Let food be your medicine and medicine be your food.* The power is in our hands; our health is in our hands. We can choose how and what we feed our body. And let's remember that our bodies and brains, like our cars, need to be powered with the right fuel to deliver the dazzling performance that we deserve.

The Results

We receive endless unsolicited testimonials from HBDers on Instagram. They write with joy about the changes they have experienced on the programme and they can't wait to share their results with us. If you join us on Instagram, you will find the friendliest and most supportive community, aka The HBD Gang, imaginable. Mostly made up of women, there are a growing number of men, too, who are understandably inspired to start HBD by their partners' transformation and rediscovered joie de vivre.

HBDers most often write to us about:

- Dramatic and sustainable weight loss. Weight loss is the motivation that draws at least 80 per cent of followers to HBD. Many, in their 40s, 50s and 60s, had all but given up with trying to lose weight and then they bump into an old friend and are astonished by their transformation. That's how most people discover HBD. Cath Weller, the face of UK fashion house Wyse, shared her HBD triumph in an article in the *Telegraph* ('Creeping Midlife Weight Gain and How to Stop it', March 2022) and that article drew lots of newbies to HBD.

- Rebalanced hormones. Women, at all stages of life, joyfully report fewer of the troublesome symptoms associated with PMT, PCOS, endometriosis, irregular periods, perimenopause and menopause.

- Improved energy. An end to mid-morning or mid-afternoon energy slumps that used to leave people desperate for tea, coffee or biscuits. We have even heard from HBDers afflicted by conditions such as chronic fatigue who have found they have a new lease of life, and for the first time in their life, once they've transitioned into Phase 4, they have enough energy to exercise. There's more energy for creativity and productivity, and for getting more enjoyment out of life.

* Better sleep. Occasionally this occurs immediately but sometimes it takes a couple of weeks once the body emerges from major detox mode. And once HBDers have adjusted to the sometimes painful shock of cold turkey – of withdrawal from sugar, alcohol, snacks, grains and processed food, and dairy – sleep is deeper and more refreshing, and people find themselves waking up earlier and naturally without an alarm and with great energy.

* A rejuvenating effect. The famous HBD glow! And not just a glow to the skin but the kind of happy glow that emanates from within. HBDers feel the benefits on the inside and see them on the outside, too.

* An improved relationship with food. That's not just a *better* relationship with food but a really *healthy* relationship with food. The treat meal is part of the key to this shift, once a week we eat whatever we fancy and realise that food is no longer the enemy. HBDers discover the foods that suit their body and that leave them feeling stronger, leaner and with vibrant energy. People write about losing the sense of diet restriction and the gratitude they feel for finding foods that suit them and their body. The healthy HBD framework makes them feel safe and in control.

* Clear skin. We've received many grateful messages from those suffering from hormonal and cystic acne, dermatitis, eczema and rosacea, whose problems gradually disappeared, along with the kilos, when following HBD.

* Less pain. A resolution of migraine headaches. Fewer aches and pains associated with IBS and certain autoimmune conditions, including rheumatoid arthritis, as well as less-painful periods that result from hormonal conditions including endometriosis.

* Less hunger! Despite the fact that HBDers are eating less food and eating it less frequently, they are astonished to find that they are also less hungry due to their rebalanced blood sugar. They feel liberated from constantly thinking about food by planning their meals in advance, knowing that they eat healthily three times a day and that the only thing they can have between meals is plain water.

A take on the Mediterranean Diet – eating the HBD way makes every calorie count towards health, energy and longevity

By following the HBD principles and rules and by using the health-giving recipes in this book, you too will feel very much better and in a short space of time. Read the HBD book, *The Human Being Diet*, and read it again so that you understand on a deep level how and why the rules work – I promise you'll find HBD easy to stick to as long as you understand the rationale behind it.

As I've mentioned, HBD builds on the famously healthy Mediterranean diet, where the focus is on liberal amounts of fruit, vegetables, legumes, nuts, seeds, olive oil and whole grains, and modest amounts of fish, seafood, dairy, poultry, eggs, and little or no red meat. But with HBD you'll find there is much less emphasis on fruit and grains and much more emphasis on vegetables and high-quality protein such as eggs and fish. And HBD also encompasses short (five-hour) fasts.

When we eat HBD nutrient-dense meals we're making every calorie that we put into our bodies count, by using wonderful mineral-dense ingredients in our meals such as seaweed. Each one of these calories can help to turn us back towards health (that's our default; the body is a self-healing organism). Your HBD experience is set to be easier and more delicious with the recipes in this cookbook, which are based on wholesome, natural and unprocessed ingredients. Eating the HBD way gives you your blueprint for life; for looking and feeling your very best, whatever your age, whatever your gender. It is a beautifully simple programme to follow and the results, as long as you stick closely to the rules, are often nothing short of miraculous.

How HBD Was Born

I've been fortunate to have some truly wonderful teachers (with wonderful names, too!) in my career, including Leo Pruimboom (Clinical Psychoneuroimmunology) and Wolf Funfack (Metabolic Balance). And these two doctors, as well as the work with my lovely clients, of course, inspired my book, *The Human Being Diet* (HBD).

I trained at the Institute of Optimum Nutrition in London and in 2004 I set up my private practice in Chelsea. I worked hard, and my practice gradually gathered momentum, and within a few years I had a waiting list. I was fortunate to have a column in the *Sunday Telegraph* magazine, *Stella*, 'My Day on a Plate', and my work featured in publications including *The Times* and *The Daily Telegraph*, as well as *Tatler* and *Vogue* magazines.

Before training in Dr Wolf's Metabolic Balance programme (MB) I was still advising people who wanted to lose weight to eat little and often. I'll never forget Wolf explaining to us during the training that if we were constantly snacking (never mind that it was mainly nuts and seeds) we were in fat-storage rather than fat-burning mode. Telling us to stop snacking felt like sacrilege, it went against everything we'd been taught. We followed his advice with trepidation and began to offer MB to our clients. And Wolf was right, it worked beautifully. At last I had a surefire tool to help people lose weight, and they lost it in the healthiest possible way. And everyone was happy.

The MB programme includes three meals a day, with weighed portions combining protein and vegetable foods, and lots of water. And no snacks! It comes with a list of foods and a meal planner, which is generated by the individual's blood test results.

To cut a long story short, Wolf sadly passed away a few years ago and the programme was changed, so I stopped using MB and started work on developing my own programme, simplifying the rules for people without compromising on the transformative results. I made the lunch and dinner portions equal sizes, and cut down on the amount of bread and fruit allowed. I increased the breakfast portions and gave people a wider choice of protein and vegetables. I also included equal weights of the protein and vegetable portions within the meals. So, in essence, as there was slightly more protein and less carbohydrate and the equal portions for lunch and dinner made it more practical to implement; it allowed for batch cooking and freezing meals, which of course saves time.

I also advised most of my clients to cut out dairy, nightshade vegetables and grains for at least the first two weeks of the programme. Like MB, HBD is a three-month programme, but unlike MB it seamlessly transitions into a new way of life to hold us in good stead forever.

I was charging top dollar and people didn't mind paying; I also did pro-bono work (as do many of my colleagues) for those who needed help but couldn't afford to pay. But the feeling that there was something wrong had been slowly dawning: people shouldn't have to be rich in order to learn how to improve their health and get more joy out of life; that information should be available to all.

I'd long been thinking of putting what I'd learned in clinic over many years into a book. A way to democratise the process of teaching people that changing the way they ate would change the way they looked and felt. A journalist contacted me in the summer of 2018 as she'd been commissioned by the *Daily Mail* to follow the programme and to write an article about her experience. And that was what spurred me on to write the book, *The Human Being Diet* (HBD). Three months later, it was done. The article was never published but that didn't matter, I'd written the book and went on to self-publish it!

FAST-FORWARD: *HBD ON INSTAGRAM*

With the first lockdown in March 2020 I had to close my in-person practice. I tried working on Zoom but it didn't work for me. I missed the intimacy and connection of meeting one-to-one in my clinic and didn't want to consult remotely. In the spring of 2020 my friend Donna Ida, aka The Jean Queen, who'd been coming to see me for nutrition advice for some time, invited me onto her Instagram channel to talk about nutrition and HBD. I hadn't engaged much with Instagram before but Donna had several thousand followers at the time and regularly invited people to talk fashion and lifestyle with her.

We did weekly Instagram lives together and the audience blossomed and grew as more and more people joined in and started to buy my book. They began to identify themselves as HBDers by using HBD in their Instagram names. And that's how the HBD Gang was born. Although Donna and I have put our regular slot on hold, I've been doing weekly lives and staying connected with the community, answering questions and offering support. I'm often joined by an HBD fan who shares their story and their experience, and the lives are always fun and fascinating.

And my dream, of democratising health and wellbeing has been realised, and almost every day I receive joyful messages from HBDers sharing how the programme has literally changed their lives. Time spent talking/messaging with HBDers on Instagram is a pleasure and I always get useful and interesting feedback. And the questions have been invaluable – thanks to the HBDers I discovered which parts of the book needed more clarification.

Changing the Way We EAT Changes the Way We FEEL

The chances are that, like 80 per cent of HBDers, you're drawn to HBD and to this cookbook because you want to lose weight. But why do so many of us struggle with our weight? Is it our genes? Or our hormones? Or a lack of willpower? Or a combination of those factors? We'll delve into the answers to those questions as we go.

It's time to dispel the myths that our old beliefs were built on. This is the end of following a depressingly restrictive diet and exercising until we're almost sick. Because we all know that calorie counting and loads of exercise just doesn't work. Exercise, and cardio exercise in particular, doesn't help us lose weight. In fact, cardio, due to the stress hormone, cortisol, can actually make it harder, almost impossible, for our bodies to burn fat.

You might well be thinking to yourself, *'What does she mean, it's not restrictive?! She's telling me not to eat bread and not to eat sugar!'* What I mean is that in following HBD we're giving our bodies absolutely the best nutrition, and absolutely what they need to function at their best. We're eating the right foods in the right ratios and at the right times; we're getting an abundance of health-giving nutrients. And, for now, we're getting rid of the empty calories and the anti-nutrients that strip our bodies of vitamins and minerals and add to inflammation and to being overweight.

Human beings are certainly not designed to be sedentary (in fact, sitting on a chair for more than an hour at a time contributes to inflammation), but neither are we designed to run back-to-back marathons. Prolonged cardio exercise, such as running, releases the stress hormones cortisol and adrenalin and pushes our nervous system into fight or flight. But what we're after with HBD is less stress and less cortisol. The antidote to high cortisol is

This is your road out of dieting and will put an end to the yo-yo effect of losing weight only to put it back on again

gentle walking. Walking, combined with HBD eating, takes us out of fight or flight and drops us into the arms of the rest and repair part of the nervous system, and that's where the magic happens.

Following this way of eating is about making new habits, ones that will stay with us forever, and breaking the old habits, habits which got us to where we are today and brought on the feeling of inspirational dissatisfaction that pushed us into trying something new. You have chanced upon a new way of eating that not only reduces your appetite and makes you feel lighter, healthier and more energetic but that also helps you to lose weight.

If you're flicking through this book and you're feeling fed up – fed up with being overweight/feeling tired/feeling generally unwell – and maybe you're reminding yourself that nothing you've tried in the past has made any lasting difference, read on. You're holding the solution in your hands: HBD to the rescue! This is your road out of dieting and will put an end to the yo-yo effect of losing weight only to put it back on again. It's time to break free!

A Look at Some of the Science, and Deep Learning

We just asked ourselves the question 'why do so many of us struggle with our weight?'. The answer is that it's often a combination of factors. But once we understand how our genes, hormones, habits and diet interact with each other to leave us feeling less than brilliant, we're empowered to make the changes that will make us feel better.

One of the most important lessons I learned about how to grow and develop into a better practitioner was thanks to Dr Leo and it relates to 'deep learning'. In my practice I was recommending big dietary/lifestyle changes to the people who came to see me. And we all know that any kind of meaningful change requires both energy and commitment; it's much easier to keep doing what we've always done.

But the people who came to see me were ready for change and ready for something new. And I owed it to them to give them the clearest possible explanations and to arm them with the knowledge they needed to make these big changes. The deeper the understanding of how and why changing XYZ in their diet would change XYZ in their health, the better my lovely clients did.

That's why there's so much explanation in my original book. When we understand the 'whys' it's easier to make the necessary changes. In fact, once we're armed with this knowledge and understanding, and as long as we're serious about wanting to change the way we feel, it becomes impossible to avoid making the necessary changes, and to help support our decision to turn back towards health in our weaker moments. And here's another hackneyed old phrase: *Knowledge is power.* Too true.

THE GENES LOAD THE GUN BUT THE ENVIRONMENT PULLS THE TRIGGER

Genes and the Art of Survival

Part of the reason for our health and weight problem harks back to our ancient past, because hardwired into our very DNA is the memory of our ancestors' constant battle to find enough food to survive. So here's a positive reframe for you: we are the survivors; we survived the drought and pestilence and famine that killed so many of our ancestors. We survivors are the ones who had the strongest immune systems as well as the wit to find the food and water we needed. And we were the ones who could eat the most when food was available, and we were exceptionally good at storing the extra calories as fat. And we still are.

Are you plagued by a sweet tooth? Let's comfort ourselves in the knowledge that we're naturally drawn to sweet tastes because our brain equates sweetness with calories and energy, and energy equates to survival. But it must be said that the more often we feed our craving for sugar, the stronger the craving becomes. I introduced you to 'The Sugar Monster' in the HBD book and explained that the more we feed the sugar monster the hungrier and the more vocal it gets. Starve it and it shrivels up and dies.

Compare that picture of our past – when food was scarce, when we had to take to our feet and use our brains to find the food we needed to survive, when we ate absolutely anything and everything that we could find – to where we are today. We're surrounded by food, by nutrient-poor ultra-processed foods (think crisps, biscuits, chocolate bars, supermarket bread, powdered soups and meals in plastic pots that we reconstitute with water) that deliver almost nothing of nutritional value except calories.

We are the caveman survivors with the genetic memory and fear of hunger, thirst and extreme cold or heat written into our genes because those are the very things that killed so many of our ancestors. Eating for pleasure, rather than for survival, and indeed eating too much, is a relatively new indulgence in our history.

So can we blame our genes for our health and weight problems? Up to a point. That glorious phrase, 'the genes load the gun but the environment pulls the trigger' means that despite the fact we might have the genes for – the susceptibility for – certain diseases, these genes may or may not be switched on (or 'expressed' in scientific lingo) and this process is known as epigenetics.

The food that we eat provides more than calories, protein, carbs and fats – it is received by our bodies as information. This information and the nutrients provided and delivered to our bodies by our food communicates with our genes and has the power to turn them on or off. The genes associated with accelerated ageing, cancer and heart disease, and many other chronic diseases, are also the genes involved with inflammation inside our bodies.

We can choose to switch on these inflammatory genes by eating lots of carbs, sugar, processed and fast food (UPFs; see below), or we can silence them by eating natural, minimally processed food, which our bodies recognise as medicine.

ULTRA-PROCESSED 'FOOD'

Bee Wilson wrote a brilliant article: 'How Ultra-Processed Food Took Over Your Shopping Basket' for the *Guardian* (February 2020) and I hear that Dr Chris Van Tulleken is writing his masterwork on this topic, which will be released in 2023. These ultra-processed foods (UPFs) are high in sugar and fat and incredibly easy to overeat. Can you imagine trying to recreate a Pringle in your kitchen? How would you flavour it? This is factory food. Part of the appeal of these UPFs is that they're cheap, much cheaper than fresh food, and they fill us up (but not for long).

Almost everything we eat and drink has been processed in some way. Milk is pasteurised; cheese and vegetables may be fermented. The differences between these kinds of processed foods and UPFs is that we can pasteurise our own milk and make our own cheese, yoghurt or sauerkraut in our own kitchen. It's when we're looking at ingredients on labels that include 'flavouring' emulsifiers, resistant corn starch, soluble corn fibre, and thickening or bulking agents that we need to think twice about the effect that these fake foods might have in our bodies. Not only on our blood sugar and insulin but also how they might affect our friendly and vitally important gut microbes. A 2019 study in the *British Medical Journal* found that even small increases of ultra-processed food in people's diets went hand in hand with higher rates of cancer, depression, heart disease and obesity.

UPFs are processed in ways that go far beyond home-style cooking or fermentation. Many of these products come with health claims printed on the packet, many of which are vegan substitutes, which are there to reassure us that this particular item is good for us. Claims such as 'high in fibre' or 'contains natural whole grains' or 'low fat' or 'gluten free' are just designed to lure us into buying these meals or snacks. We need to become avid label readers. Some products, such as lentil pasta, might be ok, but just check that label. When we see ingredients on UPF labels that we don't recognise or wouldn't use in our own kitchen at home, let's put them back on the shelf.

So just to recap, we're programmed to eat food, and to eat as much of it as possible when we get the chance. That's why we humans are still here. But now we have to turn

our backs on that old and deeply engrained programme. We're not in caveman survival mode anymore and we definitely don't need sugar or UPFs for energy. Part of the problem is that our environment has changed so rapidly that our hardwired-for-survival brain and genes haven't had a chance to keep up. Our new survival tactic is to try to ignore the food that's all around us and not to give in to temptation.

HORMONES, HUNGER AND INFLAMMATION

Whenever we eat, we activate our immune system, which results in inflammation (look up post-prandial inflammation if you'd like to know more about this). Inflammation is a normal and vital part of our immune response – the two go hand in hand.

Why is our immune system activated when we eat? Because our stomach and digestive system is our interface with the outside world and incoming food could be contaminated with bacteria (like salmonella) viruses, parasites or mould. So our immune system is activated and ready to attack all and any germs, and that helps to protect us against serious illness. The inflammation is a normal immune response to any food we eat, which is a good thing. But this also means that the more often we eat the more our immune system is activated and the more inflamed we are. Not such a good thing.

Every time we eat, and most especially when we eat carbohydrates (grains, potatoes and anything sugary or starchy) all of which break down to sugar, our body reacts with insulin (a very efficient fat-storage hormone) as well as with inflammation. High-carb and sugary foods raise our blood sugar and our need for insulin.

It is insulin's job to take the sugar out of our blood and escort it into our cells where it can be burned for energy. But when we're constantly snacking on carbs and our blood sugar is chronically high (either because our cells are becoming insulin resistant or

The more often we eat the more our immune system is activated and the more inflamed we are

our pancreas has given up on insulin production) we are also in a state of chronic inflammation. At the same time we are setting the scene for all the health issues associated with 'metabolic syndrome', which include heart problems, high blood pressure and becoming overweight, as well as insulin resistance itself.

Insulin resistance resulting from constant snacking and high-sugar foods means that glucose/sugar cannot get from our blood into our cells, which means we can't convert the food we eat into energy. And the lack of energy means we're not only tired, we are also hungry. I hope that helps to explain why snacking and overeating high-carb foods is no good for our health and why eating in this way makes us hungrier, fatter and more tired, and much less healthy, as a result.

ONE MAN'S MEAT IS ANOTHER MAN'S POISON – ANOTHER POTENTIAL CAUSE OF INFLAMMATION

Some of the foods that suit you might not suit your best friend. Some of us do fine on dairy, legumes, nightshades and grains and others definitely don't. The official advice is that it's a very bad idea to cut out whole food groups and that if we do so our health is likely to suffer. That's just not true. It's equally untrue to assert that we need dairy foods for calcium and that these foods are vital for bone health. We don't and they're not.

Let's consider the advice to eat lots of whole grains for health. When I started my training in nutrition, we were taught that oats provided a healthy and nutritious breakfast. So I duly started eating oats in the morning – sometimes with milk or water, sometimes with fruit or seeds. And when I ate oats for breakfast, I found that either I was starving hungry an hour or so later or I felt so tired I had to go back to bed. But other people find they do brilliantly well on oats for breakfast – it keeps them going, and with great energy, for hours. Different strokes for different folks, but oats aren't for me.

I also discovered, through one of our homework assignments, that wheat did not suit me. We were instructed to cut out all wheat for at least 14 days and to make a note of any changes we felt when we reintroduced it. I was a diligent student; I read labels and avoided it 100 per cent. I reintroduced it on day 15 and a couple of hours later I was distinctly bloated and sluggish. But when I woke up the next day I felt really bad, like the end of the world had come; I had no energy and didn't want to get up. I didn't want to do anything; I was enveloped in a black cloud of depression.

It took me a few days to recover, and I vowed never to eat wheat again. Fast-forward about 10 weeks and the perennial black bags under my eyes had disappeared. Not only that but I also felt much more energetic and much cheerier.

Similarly, many HBDers have found that when they eliminate dairy completely, as recommended for the first 16 days of the programme, they're shocked at how their body reacts when they reintroduce it. Horrendous stomach or joint pain, nausea, bloating, facial spots and blocked sinuses, to name a few. And these HBDers had no idea that dairy was not good for them either.

That's what I mean by listening to the feedback from our bodies. None of us know whether or not a specific food (dairy and grains are the usual suspects) suits us unless we completely eliminate it for some time before trying it again. That's why we recommend you take out dairy as well as nightshades (and grains are automatically out) for the first two phases and test these foods on yourself in Phase 3.

NO SUBSTITUTIONS!

I'm often asked questions such as: can I have peas instead of lentils or can I have kefir instead of yoghurt, or can I have oat milk or almond milk in my coffee? Or can I have dairy milk in my coffee if I'm having yoghurt for breakfast as it's the same protein, or can I have chia seeds instead of sunflower and pumpkin seeds, or can I have half my apple with breakfast and half with lunch? And the answer is always no! The foods listed provide the most bang for the buck – they contain the highest-value nutrients per calorie and, by trial and error, have been shown to deliver the best results.

Introducing the HBD Programme and *The HBD Cookbook*

My aim with *The Human Being Diet* was to give you a blueprint for feasting and fasting your way to feeling, looking and being your best, whether you want to lose weight or not. HBDers have long been asking for a cookbook to help them make following HBD more enjoyable, and here it is at last. I'm so proud to offer this cookbook, which is designed to help make your HBD experience easier and more delicious. I hope it will inspire you to create your own recipes, too, by following the weights and the combinations of foods found here. And because we recommend that you repeat the first two phases of the HBD programme once a year as a Reset, you will have this book at your side for some time to come!

You can follow the HBD principles and use these delicious recipes to help you lose weight and to improve your energy and overall health, but to understand the programme, and to call yourself an HBDer and become part of the HBD Gang, you need to read my original book. The book (it's an easy read, I promise) explains in detail why the rules are in place and the science behind them. And as you read earlier, when we understand how and why the food we eat has a direct effect on how we feel, it makes it much easier to follow the rules. In fact, when we understand the rules and how and why they work, it becomes impossible NOT to follow them if we really care about our health and how we feel.

WHAT IS THE HBD PROGRAMME AND HOW DO WE DO IT?

To recap: HBD is a programme that addresses blood sugar and insulin levels, which has a knock-on effect of rebalancing hormones and reducing overall inflammation, with a side-effect of weight loss. When followed correctly and according to the rules, HBD teaches us about which foods suit us personally as biochemically unique individuals, and often which ones don't; it resets our relationship with our food, and how and when we eat it.

We're going to be eating three meals a day, with a minimum of five hours without food between each meal and absolutely no snacking. No tea or coffee between meals either, just water. And nothing in the water other than unsweetened and unflavoured electrolytes. Unsweetened black tea and coffee is allowed, but only with meals.

The meals are made up of fresh and wholesome ingredients that strictly include one type of protein, a mixture of vegetables and, as an option, one type of fruit per meal (but an apple a day is compulsory). Once you're into Phase 3 – from day 17 onwards – you can have a weekly treat meal. And that's compulsory too, not just because it's fun but because, believe it or not, it's vital for fat burning and for overall health. Read more about that in the introduction to Phase 3.

Why three meals a day? Why not have long overnight fasts? The key to the success of this programme is in re-establishing stable and healthy blood sugar levels and the right combination of foods, eaten three times a day at the same time achieves just that – the body's expectations follow a natural rhythm. When our blood sugar is stable, when we're no longer experiencing a roller coaster of blood sugar highs and lows, both our energy and mood stabilise and improve. Not only that, but after the first few days we're no longer hungry and we're no longer plagued by cravings.

The lack of hunger surprises HBDers because they're eating much smaller portions than they're used to and they're eating less frequently. When you first start it's completely normal to look at your plate in disbelief and you'll probably find yourself thinking gloomily, 'is that it?' Rest assured, nearly all HBDers feel the same way when they start, and they are astonished to find that they adapt to the smaller portions within a few days.

WEIGHT LOSS – HOW MUCH WEIGHT WILL YOU LOSE?

It's impossible to predict how much weight you are likely to lose each week – some weeks you might lose a couple of kilos and some weeks nothing. But looking back at your starting weight, once you get to the end of the three months, you're likely to find you've lost an average of 1.5kg per week. But a lot depends on your starting weight – normally the more there is to lose, the faster it comes off.

Getting Ready to Start: THE MINDSET

You've read this book and have an idea of how and why HBD works, but perfectly normal questions like these might be running through your mind: How hard will it be? How will I feel? Will it work for me? Will I be able to stick to it?

Grab a pen and paper, or your phone, and start writing down the things that drew you to HBD in the first place. Answering these questions might help to get you started:

- What did you hear about HBD that resonated with you and first got you interested?
- What do you want to change? Something about the way you feel? Or the way you look?
- Are you interested in HBD because there are niggling health issues you want to address?
- Is it because you'd like more energy? Or better sleep? Or less pain?
- Were you inspired to start by a friend?
- Are you looking for more of a sense of control in your life?
- Have you got a long-awaited holiday coming up? Or a wedding anniversary or a milestone birthday?
- Are you thinking about your long-term health with your children in mind? Or hoping to start a family one day?

Start writing down your reasons for wanting to start – the inspirational dissatisfaction that made you want to change. This is for your eyes only, so make it as personal to you as you possibly can. It might be an emotional experience – that's normal; just go with it and take time over it. It's important to make your reasons real and valid for you.

If your biggest reason is that you want to fit into the clothes that take up half your wardrobe but they no longer fit, ask yourself why this matters so much. How it will make you feel when you slip into them again; what would this mean to you?

When you're really clear about your reasons for starting HBD and you've acknowledged their importance to you, you will begin the programme with the resolve and determination to see this through. Keep a record of your 'whys' and revisit them regularly. This, and following the rules, of course, is key to your success. Join the HBD Gang on Instagram @petronellaravenshear or buddy up with a friend or two for support and to help keep you accountable.

Look at your diary and find the clearest 16 days you can, bearing in mind there's never a perfect moment to start but starting will make the moment perfect, then put it in your diary and commit!

EGGS

Eat the white and yolk; avoid egg white omelettes! Why? Not only because an egg white omelette is a depressing 'diet food' and the yolk is avoided by people who are needlessly frightened of fat, but because a whole egg is one of the best protein sources on earth. And it's the combination of the white and the yolk that give it a perfect balance of amino acids, which make up a protein. If you had egg or eggs for breakfast do not have eggs again the same day. Variety is key on HBD.

Following the Programme – How to Do It

The programme is simple to follow. It's split into four phases over 12 weeks and there are 10 golden rules. And as long as you stick closely to the rules, success is yours. Prepare to feel fabulous! Here's a summary of the four phases – this is just a brief introduction, you'll find more details at the beginning of each chapter.

If you're serious about HBD and you want the best possible results, go pure! Don't be tempted to cherry-pick your way through the rules by making a kind of hybrid between HBD and any other programmes. If you are following some of the HBD rules and incorporating parts of other eating programmes/diets, it won't work, and you won't get the results you deserve.

Phase 1: Preparation – two days of vegetables only. No oil, fruit, sugar, protein, grains, alcohol or dairy.

Phase 2: Reset – 14 days of three meals a day. No oil, sugar, grains or alcohol and for best results also avoid dairy and nightshade vegetables.

Phase 3: Burn – 10 weeks minimum. Olive oil is reintroduced along with a weekly treat meal.

Phase 4: Forever – Forever!

Here Are Your 10 Golden Rules

1. Eat three meals a day and fast for five hours, with plain water only between meals.
2. Begin each meal with a couple of bites of protein and eat just one type of protein per meal.
3. No oil and no alcohol for the first 16 days.
4. No wheat or any other grains for the first 16 days.
5. No cardio exercise for at least the first 16 days.
6. Drink the right amount of water (about 35ml per kilo of body weight).
7. Eat one apple a day with a meal and only eat one type of fruit per meal.
8. Don't eat for longer than one hour (except at weekly treat meals).
9. Finish eating by 9pm.
10. No sugar (and no honey or fake sugar) except at treat meals.

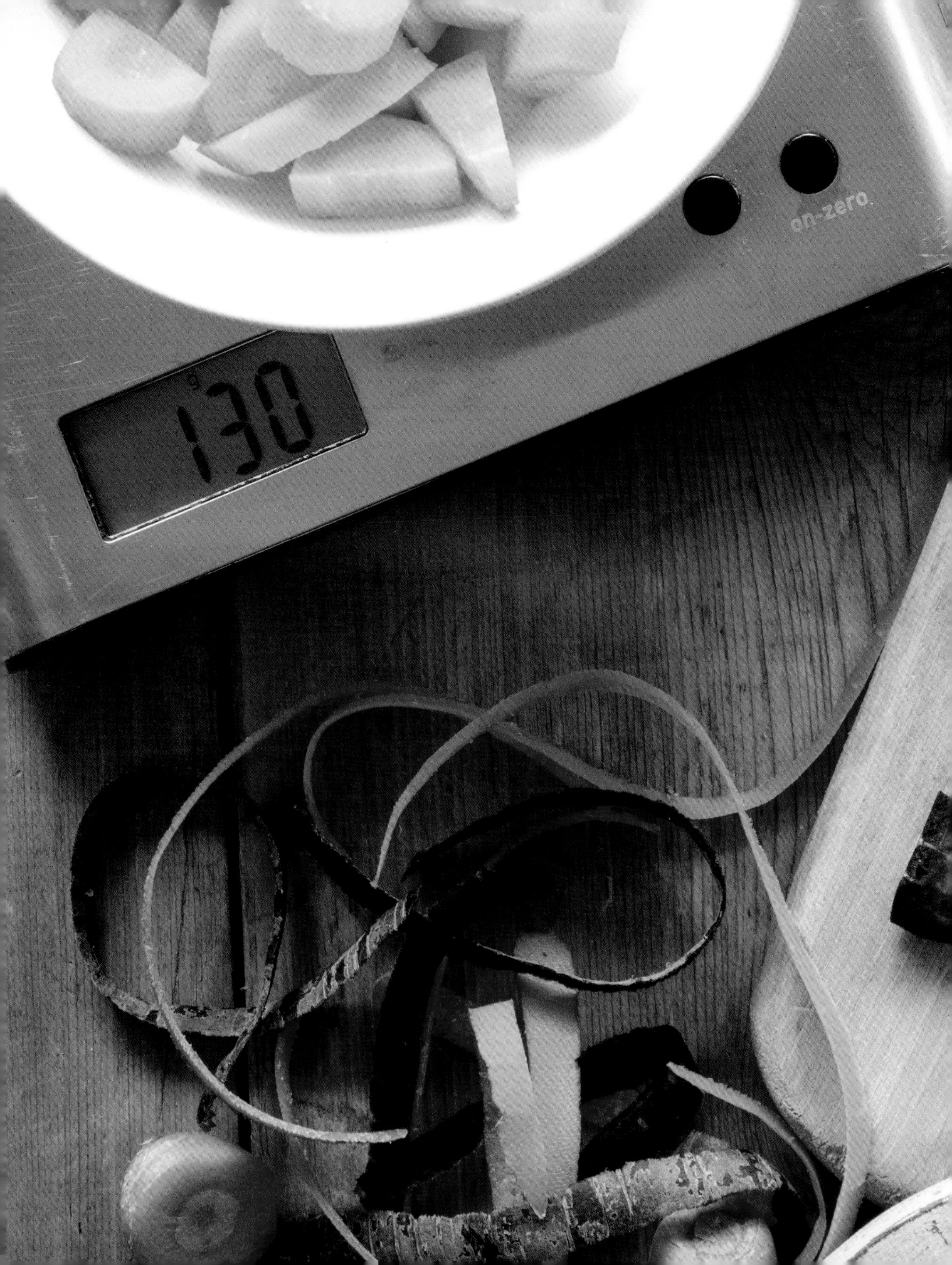
130
on-zero

RULE 1:

EAT THREE MEALS A DAY AND FAST FOR FIVE HOURS, WITH PLAIN WATER ONLY BETWEEN MEALS

This is the most important rule of all. Two of the many hormones involved in inflammation and weight gain or weight loss are insulin and glucagon. Insulin is our fat-storage hormone and glucagon is our fat-burning hormone. When we're fasting between meals, we have lower insulin and higher glucagon levels, which results in more fat burning and less inflammation, and that's exactly what we want.

Eating three HBD meals a day, always combining protein with carbohydrates (mainly vegetables), results in steady blood sugar and energy levels and less hunger through the day. The five-hour fast between each meal gives the digestive system a rest, and it also helps to reset our internal clocks and our circadian rhythm, which helps to rebalance all our hormones.

Bear this in mind: any time we snack or drink anything other than plain water between meals, we're sabotaging our body's attempt to burn fat and reduce inflammation. Drinking anything other than plain water between meals, whether it contains calories or not, can potentially impact our blood sugar and therefore our insulin levels. Tea or coffee with milk (milk contains a sugar called lactose and a protein called casein) acts like a mini-meal, which means there's a small rise in blood sugar and fat burning stops.

Black coffee and tea contain caffeine not calories. Caffeine affects the nervous system and puts us into mild fight or flight; sugar is released into the bloodstream and fat burning stops. And that's why some diabetes experts forbid their patients to drink coffee, even decaf. But for most healthy people it seems that a cup or two of black tea or coffee with or just after a meal has a minimal effect on blood sugar and doesn't interfere with fat burning.

If you were a snacker, or a frequent tea or coffee drinker, you've inadvertently trained your brain and body to expect food/mini meals every two hours or so. It's going to take a few days for blood sugar to stabilise again and while you're making your great new habit of fasting between meals you'll probably feel hungry and angry – hangry – and maybe light-headed. But stick with it, keep the faith, your body will adapt and pretty soon you will be making the vital change from carb-burner to fat-burner. Your body will adapt to the lack of calories coming in by reducing inflammation and burning stored fat for energy.

RULE 2:
BEGIN EACH MEAL WITH A COUPLE OF BITES OF PROTEIN AND EAT ONE TYPE OF PROTEIN PER MEAL

If our body's got used to dealing with lots of carbohydrates, including sugar, fruit and grains, the pancreas gets trigger-happy and with the first sign of a carbohydrate (even a vegetable) it ramps up insulin production. We want lower insulin levels, which is why we begin each meal with a couple of bites of protein.

Breakfasting on good-quality protein (eggs, fish, chicken or tofu) not only keeps us fuller for longer but has also been shown to keep our blood sugar levels stable for longer. And getting good protein has an effect on another hormone, leptin. We can become insulin resistant and we can also become leptin resistant, which means we're constantly hungry. Eating protein within an hour of waking helps us overcome leptin resistance so we're no longer hungry.

Because we're eating smaller meals it's vital that we maximise protein absorption from them. Combining two or more different types of protein within a meal can result in our body absorbing less protein. Eggs, for example, contain a perfect balance of amino acids to make a complete protein, but if, for example, we combine eggs with salmon we actually absorb less protein. The same goes for mixing any other protein foods together, including soy.

And speaking of protein absorption, we can maximise the digestion and absorption of all the nutrients in a meal by relaxing before sitting down to eat and by focusing on taking our time and properly chewing our food. Because as long as we're relaxed our digestive system kicks in long before the food arrives in our stomach. It begins with the aromas and anticipation of the food we're preparing. Digestive enzymes in saliva in the mouth get to work and that, along with more chewing, improves our digestion and absorption no end.

RULE 3:
NO OIL AND NO ALCOHOL FOR THE FIRST 16 DAYS

This is a liver holiday and time to give this overworked organ some TLC. The liver does incredibly important work for us, including fat burning and keeping blood sugar levels stable. But we often mistreat it. Not only with alcohol and deep-fried or fatty food but also with too much fruit. Time to give the liver some love with light nutritious meals, herbs and green vegetables.

RULE 4:
NO WHEAT AND NO GRAINS FOR THE FIRST 16 DAYS

Contrary to conventional wisdom, we don't need to eat grains at all, wholegrain or not. In fact, most of us find we feel much better without them. Gluten grains (wheat, rye and barley) are particularly problematic, but wheat is the one that causes many of us trouble such as bloating, skin problems and so on. In Phase 3 you can try rye bread – up to 100g in addition to your protein and vegetables – once or twice a week. See how it makes you feel. You could try bread or pasta (or cake or pastry) at a treat meal and keep tabs on how you react. But for now, in Phases 1 and 2, stay off all grains.

RULE 5:
NO CARDIO EXERCISE FOR AT LEAST THE FIRST 16 DAYS

This will come as welcome news for some of us but for others it might be a rule that's tempting to break or ignore. Cardio exercise and HBD are mutually exclusive; it's a case of choosing one or the other for now, you can't do both. But this isn't forever; once you're well into Phase 3 and at your happy weight, you can take up cardio again. The same goes for weight training. In the meantime, stick to walking, stretching and gentle yoga.

If you're already fit and work out regularly, continue with a light version of your normal regime. A 'light version' means that you can keep up a normal conversation and you're not out of breath. But if you don't already work out this is definitely not the time to start. Avoid any kind of exercise that makes your heart race (such as spinning, skipping and running) and which leaves you sweaty and out of breath.

'I have two doctors, my left leg and my right. When body and mind are out of gear (and those twin parts of me live at such close quarters that the one always catches melancholy from the other) I know that I shall have only to call in my doctors and I shall be well again.'

G. M. Trevelyan

Why no cardio? Because cardio pushes our nervous system into fight or flight – it's a stressor. Think about the two parts of the nervous system as the accelerator and brakes of a car. The accelerator, which increases our stress hormones adrenalin and cortisol, is the sympathetic system, aka flight or fight and GO! The brakes are parasympathetic, aka rest and repair and STOP! By taking our foot off the accelerator and using the brake pedal more often we're allowing our nervous system to relax.

The effect of restoring normal insulin and blood sugar levels, via HBD, allows our hormones to regain their natural rhythm. Cortisol, like all our other hormones, follows a circadian rhythm; when we're healthy, our cortisol levels are naturally higher in the morning and lower in the evening. But when cortisol rhythms are out of whack, all our other hormones are impacted, leading to wide-ranging effects on weight, sleep, mood, energy, skin and digestion. It also impacts periods and female hormone balance.

If we feel depleted, exhausted and/or stressed, it means we've lost our rhythm and our ability to adapt to stressors, be they physiological or psychological. That well-known feeling of 'tired but wired' – too wired at night to sleep well and too tired in the day to concentrate and function as we should – is down to dysregulated cortisol rhythms. It's a sign that we've lost our natural rhythm.

So, to sum up, of course exercise is good for us, that goes without saying. But for now, while we're re-finding balance and encouraging our bodies to reduce inflammation and get on with fat burning, we need to avoid exercise that pushes us into high-cortisol flight or fight mode. We want lower cortisol levels and to spend much more time in parasympathetic mode. And the best exercise to lower cortisol levels? Walking!

JOURNALLING AND KEEPING A FOOD DIARY

The most successful HBDers not only decided from day 1 that they were committed to giving HBD their all, but many of them also kept (and some still do) a food and symptom diary. This is an ongoing daily food diary combined with a journal to record what you eat and drink, and how it makes you feel, both physically and mentally/emotionally.

It sounds like a chore, but its value and importance cannot be overstated. One HBDer found that her weight loss stalled in the middle of Phase 3 and she couldn't work out why until she looked at her diary. She noticed she'd got into a rut of eating pretty much the same foods every day, and specifically the same vegetables. So, she shook things up and introduced new vegetables and sure enough the weight started coming off again.

It's the best way to track reactions to food that you've newly introduced in Phase 3 and in treat meals too. Maybe the first time you have wheat and you wake up the next day feeling bloated and sluggish, you put it down to coincidence. But if the same thing happens the next week, I hope you'll stay off wheat for now and maybe try it again in a few weeks.

Another Phase 4 HBDer who's still tracking her food worked out that she's fine with yoghurt and cheese but that when she has cream she wakes up the next day with blocked sinuses. A clear message to her from her body saying 'I don't like this!' She doesn't like the sensation so no longer has cream.

RULE 6:
DRINK THE RIGHT AMOUNT OF WATER

This is right up there with Rule 1 – water is nature's anti-inflammatory and it's absolutely vital that you drink enough of it to energise and detox your body. Drink 500ml water before breakfast and before you do anything else in the morning. Aim for at least another 1.5 litres before lunch. Drink water when you're hungry, when you're bored, when you're tired, when you're angry. It sounds ridiculous, but truly, water is miraculous stuff. Because dehydration killed so many of our ancestors, our bodies react to a lack of water by slowing our metabolic rate, which means less detoxing and less fat burning.

RULE 7:
EAT ONE APPLE A DAY WITH A MEAL AND ONLY EAT ONE TYPE OF FRUIT PER MEAL

Apples are an excellent source of the detoxifying fibre pectin, as well as antioxidants including quercetin, which is a natural antihistamine. So, eat one apple once a day with one of your meals. If you're wondering if you can substitute a glass of apple juice, even if it's fresh, for one apple, the answer is no! If you juice your apple, you're missing out on the wonderful fibre that it contains.

Oral allergy syndrome (OAS) occurs in some people when they eat apples (and some other fruit and raw nuts), which results in itchy mouth, ears and throat as well swollen lips or hives. If you suffer with OAS, and as long as your doctor hasn't advised you to avoid apples, try eating them in a different way, in this order: baking them in the oven or poaching in water. If you can tolerate eating them like that for a couple of weeks, move on to trying blended apple. Progress from there to grating it (it's delicious with seeds or walnuts and some cinnamon for breakfast) and by then you might be ready to try cutting it into slices and eating without any adverse effects. But if you can't find any way to make it edible for you, leave it out altogether.

RULE 8:
DON'T EAT FOR LONGER THAN ONE HOUR (EXCEPT AT WEEKLY TREAT MEALS IN PHASES 3 AND 4)

This is to avoid the grazing effect. Unless we've got our five-hour fast between meals, our insulin isn't low enough for us to burn fat. Start the five-hour fast from when you finish your meal and not from when you start it. And remember that if you're having black tea or coffee with your meal that the five-hour fast doesn't begin until you've finished that too.

RULE 9:
FINISH EATING BY 9PM

Remember that for less inflammation and for more fat burning we need lower insulin and higher glucagon and as long as we finish eating before 9pm (8pm would be even better) our insulin levels will be at their lowest overnight; that's why we're in fat-burning mode while we sleep.

RULE 10:
NO SUGAR (AND NO HONEY OR FAKE SUGAR) EXCEPT AT TREAT MEALS

The reason for the no sugar rule is, I hope, a bit of a no-brainer. All sugary and sweet foods, including honey, are easily converted into fat. Sugar is pro-inflammatory and ageing. It strips our bodies of the nutrients, including B vitamins, magnesium and chromium, that we need for energy and immune resilience.

Sugar certainly gives us a quick energy hit but ultimately it leaves us more exhausted and looking for our next 'fix'; the more sugary food we eat the more we want, which perpetuates the high/low blood sugar and energy roller coaster.

You've possibly been surprised by articles suggesting that those who drink zero calorie fizzy drinks, are prone to weight gain. Why would this be? There are three plausible/possible reasons:

1. Sweet tastes make us hungry for more.
2. Our brain senses the sweet taste and expects energy to be on the way in; but when the energy doesn't arrive we're hungry and driven to eat.
3. Fake sweeteners decimate our friendly gut microbes in a similar way to antibiotics (and farmers have discovered that feeding animals antibiotics puts more weight on them, which means they get more money for them at market).

WHY EAT ORGANIC?

The short answer to this question is that organic is better for humans and animals and better for the environment too. The US-based EWG (Environmental Working Group) publishes their annual and invaluable guide 'The Dirty Dozen' and 'The Clean Fifteen' on their website (www.ewg.org/foodnews/dirty-dozen.php). Do read up on their research and bear in mind that when we eat organic produce we avoid the cocktail of pesticide residues that's served up with more than 70 per cent of non-organic fruit and veg. And, as EWG states, these pesticides are not only toxic to humans and animals but many of them are also toxic, i.e. they kill, our precious bees. When it comes to animal foods, check out the UK Soil Association's website. Organic farming ensures that animals are truly free-range and that they have the best possible living conditions. Yes, organic is more expensive but I believe we owe it to ourselves and to our planet to eat organic wherever and whenever possible. www.soilassociation.org/take-action/organic-living/why-organic/better-for-animals/

WATER!

Your best HBD friend. Water takes the edge off hunger and keeps our metabolic rate ticking over nicely, which keeps us in fat-burning mode. It also helps to flush out toxins and gives us energy. Drink 500ml first thing in the morning and drink at least another 1.5 litres before lunch. Remember that a rough guide is 35ml water per kilo of body weight. If you weigh 60kg you need about 2 litres of water and if you weigh 80kg you need nearly 3 litres.

Drinking most of your water before lunch means you won't suddenly realise at 4pm that you haven't had enough and end up bulk drinking in the late afternoon/early evening. You don't want to be up all night nipping in and out of the loo; night time is for sleeping and fat burning.

Ideally, use a filter jug or another form of filtration – it makes the water taste better (with less 'scum' in your tea or coffee) and it removes heavy metals such as lead and mercury. We often hear that London drinking water is the purest, and while it may be true that once it's been filtered and treated it's exceptionally pure, think about all the old lead pipes it travels through before it reaches our taps.

Whenever you're hungry, bored, angry or tired, drink water. Once you start to drink more water you're likely to realise that you need more and more of it – all those times you thought you were hungry, you were thirsty! Avoid adding lemon or mint or cucumber or anything else to water between meals, but you might find that unflavoured electrolyte drops help it slip down. If you're serious about getting the results you want from HBD and especially with regards to inflammation, weight loss and detoxification, drink your water.

Top Tips for Starting

- Revisit your reasons – what was it that drew you to HBD? If you haven't already written these down, in as much detail as possible, do that now (see The Mindset on page 33).
- Take your measurements because sometimes although we've burnt fat we've also gained muscle and the scales don't reflect that.
- Weigh yourself (this is optional but most of us find it motivating) and weigh yourself once a week going forwards – the morning before your treat meal perhaps.
- Keep referring back to the book – if a food isn't listed it's not allowed.
- Remind yourself that if you stick like a limpet to the programme, you'll get the incredible results that you want and deserve.
- Make sure you have the right equipment and ingredients to start your journey.

EQUIPMENT

Here are some of my recommendations.

Tape measure –so you can take your vital statistics. Measure around your waist, around the top of your thigh and the widest part of your hips on day one and repeat every couple of weeks.

A selection of small bowls and plates – buy yourself some appealing bowls and plates; fooling the eye fools the stomach.

Digital scales – to accurately weigh out your ingredients. Always keep a spare battery on standby.

Spiraliser – although not essential, these are great for making vegetable noodles

Walter filter jug – you'll be drinking a lot more water so invest in one of these for better-tasting, purer water. You'll also be able to keep track of your water intake more easily.

Epsom salts – not exactly equipment, but a must for your first day on HBD! Westlab Reviving Epsom Salts are perfect and safe to consume (see page 54) and come in large bags for a more economical option (you can also use these for a relaxing bath).

INGREDIENTS

Always buy organic whenever you can (see page 47). Here are a few of my favourite food brands:

Daylesford – yes, it's expensive but remember how much money you're saving on cappuccinos and snacks. Splurge on organic produce for at least the first 16 days if you possibly can; your body and your friendly gut microbes will be forever grateful. Daylesford are suppliers of highest quality protein foods and seasonal varieties of fresh and fermented vegetables, including salad bags and herbs straight from their own kitchen gardens. Their clear Organic Chicken Bone Broth can be used in place of oil for stir-fried meals in Phase 2 and doesn't count as mixing your proteins. www.daylesford.com

Willy's Organic Apple Cider Vinegar – made by Willy, a veritable pioneer in biodynamic probiotic farming, with apples from his ancient orchards. Try it in a wine glass with ice and sparkling water with your HBD meals and feel it doing you good as it slips down. Only use the original unflavoured ACV with the blue label. www.willysacv.com

Bold Bean Co – incredibly delicious chickpeas and butterbeans. wwwboldbeanco.com

Abel & Cole – the original pioneers of organic fruit and vegetable box deliveries with salad and herbs. www.abelandcole.co.uk

Hurly Burly Foods – all organic; the sauerkraut specialists providing mixes of vegetables, spices and herbs to nurture our gut microbes. Note: start with very small amounts of sauerkraut to avoid wind/tummy upsets. www.hurlyburlyfoods.com

Kape – the creamiest and crunchiest activated walnuts, delicious served with your daily apple for breakfast. www.kape.co.uk

Hunter & Gather – suppliers of healthy kitchen staples; use their classic egg-free mayo as an occasional treat in Phase 3 and as it's egg-free you won't be mixing your proteins if you serve a dollop on your fish or chicken. www.hunterandgatherfoods.com

Pukka – the best sustainably grown organic herb teas available. www.pukkaherbs.com

Steenbergs – offers a range of organic and Fairtrade herbs and spices. www.steenbergs.co.uk

PHASE 1 Preparation

Phase 1: Preparation

- **Two days of vegetables only**
- **No oil, fruit, sugar, protein, grains, alcohol or dairy**

On the first day, 30 minutes or longer before breakfast, dissolve 3 teaspoons of Epsom salts into half a mug of hot water. The salts don't taste great but as long as you have a good clear out on the first day (they have a laxative effect) you won't need to repeat the dose. Follow the Epsom salts with lots of plain water and stay near the loo! If, which is unusual, the salts don't work on day one, repeat the process 30 minutes before breakfast on day two. If you still don't go, repeat 30 minutes or longer before lunch on day two. The purpose of the Epsom salts is to minimise detox symptoms including headaches. But if you do get a bad headache (rare) and need paracetamol, take it.

NIGHTSHADES

Avoid these absolutely from day one if you have any inflammatory or autoimmune conditions including rheumatoid arthritis, osteoporosis, psoriasis or IBS. Nightshades, especially when we have underlying health conditions, are apt to fan the fire of inflammation. How would you know you might have inflammation even without a medical diagnosis? Does anything hurt? That's a giveaway. HBD purists eliminate nightshades for the first two phases and reintroduce them in Phase 3 and see how their body reacts. Nightshade vegetables include peppers, chilli peppers (and Tabasco), aubergine, tomato and potato.

VEGETABLES, VEGETABLES, VEGETABLES

For the first two days it's vegetables only – no fruit, sugar, oil, alcohol, grains, pulses/legumes (including soy, peas and mangetout), potato, sweet potato, sweetcorn, protein or dairy. Just glorious health-giving vegetables and salad. Tomato and avocado, although technically fruits, can also be included. Unlike in Phases 2 and 3 there's no need to weigh your Phase 1 meals and no need for the five-hour fast between your meals – this is a detox moment, like lots of little brooms clearing out your system and feeding your microbiome. These two days are about celebrating above-ground and below-sea vegetables. The plan is adaptable and flexible, so if you get a weekly veg box delivery, you can experiment with vegetables you haven't tried before. Do include other veg such as seaweed, which is chock full of minerals. Be brave and eat that rainbow!

In Phase 1, aim for around 500g of vegetables per meal, including at least 3 types of veg

SLEEP

Believe it or not we burn more fat when we're asleep than when we're awake – it's when our body gets on with repairing damage incurred from the day before and when our immune system is most active. It's when the brain is in detox mode – clearing out debris that keeps our thinking sharp and clear the next day. Don't stint on it!

Aim to be in bed with the light out by 10pm and you're likely to wake up feeling cheerful and clear-headed by about 6am. Switch off your blue-light-emitting phone/iPad/TV at least an hour before bed and wear blue-light-blocking glasses in the winter when it gets dark in the late afternoon. Why? Because blue light is akin to daylight and keeps us awake by preventing the release of our sleep hormone, melatonin.

FLAVOUR

Salt and black pepper and fresh, dried or frozen herbs – including rosemary, basil, coriander, thyme, sage, etc., plus garlic, turmeric and ginger as well as apple cider vinegar – can be added freely to your meals. As can fresh liquid vegetable or chicken stock (if you're making homemade chicken stock make sure you skim off any fat – easily done when the stock/broth has been refrigerated for a few hours). No stock cubes or powdered bouillon!

If you're avoiding nightshades, which include chilli peppers, try annatto, which is a red-orange seed from Mexico – it's tangy and tastes sweet and smoky.

DRINKS

Unlike in Phases 2 and 3, black and unsweetened coffee and tea (or green or herbal tea) can be drunk freely throughout the day because we're not yet thinking about our blood sugar. Other than that, just drink plain water, flat or fizzy. No coconut water until Phase 3 as part of a treat meal or with meals in Phase 4 (it breaks the fast between meals).

POOPING!

A common side-effect of following HBD is a change of bowel movements. Some people notice that for the first time in their life they become 'regular' while others seize up, and aside from the odd rabbit dropping there's not much action. Pooping is an all-important part of the detox but constipation sometimes occurs as a side-effect of eliminating wheat and sugar. If you're drinking your all-important morning water and walking regularly but still not going, try supplementing with magnesium, either as citrate or glycinate, 300–400mg with breakfast and another 300–400mg with food in the evening. Vitamin C can help, too. If you're still 'stuck' try Cascara Compound (not plain cascara), which is an old naturopathic herbal formula and has a toning and stimulating effect on the bowels.

N.B. Phase 1 is just the first part of the detox, which continues into Phases 2 and 3.

A note for old-time/experienced HBDers: you will notice that sweet potato and beetroot no longer feature in Phase 1. Feel free to continue to include these in Phase 1 of your annual Resets. The reason they're now omitted is because HBD newbies assumed that because they were included in Phase 1 they were ok to eat throughout Phase 2. Sweet potato and beetroot can be introduced from Phase 3 onwards.

THE SUGAR MONSTER

Any time you're tempted to eat something sweet (even fruit) remind yourself that the problem with sugar is worse and runs deeper than it simply being a source of empty calories. Sugar is an 'anti-nutrient' – it depletes our bodies of vitamin C, magnesium, chromium and B vitamins, to name but a few. And these nutrients are the very ones that we need for creating energy in our cells. So it's absolutely true to say that sugar makes us tired and even depressed. And it makes our skin grey and wrinkly. And it makes us puffier and fatter and maybe even spottier too. The longer you avoid it, the easier it is to kill the sugar monster. Remember that adding apple cider vinegar to water with meals can also help to curb the cravings.

Quick and easy meals to get you started

BREAKFAST, LUNCH OR DINNER

Half an avocado with four cherry tomatoes, salad leaves and cress, basil leaves, apple cider vinegar, salt and pepper.

• • • • • •

Any kind of mushrooms sautéed in stock with garlic; garnish with parsley and serve with mixed green leaves and three or four cherry tomatoes.

• • • • • •

Vegetable soup – made with fresh or supermarket stock and add as many different herbs and vegetables as possible.

PHASE 1 CORE ACTIONS

Aim for around 500g of vegetables per meal and ideally at least 3 types of veg.

This phase includes vegetables from the list below plus stock, herbs, apple cider vinegar and spices only.

Permitted Phase 1 Foods – if a food is not listed here, it's not permitted!

Apple cider vinegar
Artichokes, fresh (hearts bottled in brine not oil are ok)
Asparagus
Avocado
Aubergine *
Bell peppers *
Black coffee
Black pepper/sea salt
Black tea/green tea/herb tea
Broccoli
Cabbage
Carrots
Cauliflower
Celery
Chicory
Chillies*
Chinese leaves
Courgettes/zucchini/marrow
Cress
Cucumber
Fennel
Garlic
Ginger
Green beans
Herbs – fresh, frozen or dried

Permitted Phase 1 Foods – continued

Kale
Kohlrabi
Leeks
Lettuce
Mushrooms
Okra
Onions
Pak choi
Paprika*
Pumpkin
Radishes
Rocket
Salad leaves – all
Seaweed
Shallots
Spices – no mixes that contain anything other than herbs/spices (e.g. no sesame seeds), and no oils/additives
Spinach
Spring onions
Sprouts
Squash – all types
Tomatoes*
Vegetable stock – homemade or liquid from supermarket
Water

*Nightshade vegetables are marked with an asterisk. Recipes that include nightshade vegetables are marked with and should be avoided for HBD purists and anyone with inflammation.

Banned Phase 1 Foods

- No alcohol
- No fruit
- No grains
- No sweetcorn
- No dairy (animal or plant), including cheese, milk, yoghurt
- No sugar or sweeteners, including honey
- No pulses (e.g. peas, broad/navy/cannellini/kidney/edamame beans, soy, lentils, chickpeas)
- No protein (eggs, fish, shellfish, poultry, soy, meat)
- No oil
- No tamari , soy sauce or miso
- No nuts or seeds
- No herbs or spices other than fresh, dried or frozen
- No stock cubes – use fresh liquid stock only
- No potato or sweet potato
- No beetroot
- No fruit tea
- No juices! Whole fresh veg only

Recipes including tamari are marked with and best avoided by HBD purists until Phase 3. There's nothing wrong with tamari, but better to allow the natural flavours to shine through!

PHASE 1 RECIPES

Zingy Aubergine & Tomato Curry

You can subtly change the flavour and aroma of this colourful curry by tweaking the spices and flavourings.

SERVES 2
PREP: 15 MINUTES
COOK: 30–35 MINUTES

2 onions, chopped
3 garlic cloves, crushed
2.5cm piece of fresh ginger, peeled and diced
300ml good-quality vegetable stock
2 tsp ground turmeric
1 tsp ground cinnamon
1 tsp ground coriander
½ tsp ground cumin
3 vine tomatoes*, quartered or cut into chunks
1 medium aubergine*, trimmed and cubed
250g fresh spinach leaves, washed, trimmed and shredded
a handful of coriander, chopped
sea salt and freshly ground black pepper

1. Put the onions, garlic, ginger and stock into a large saucepan over a high heat. Cover the pan, bring the stock to the boil and then boil for 10 minutes. Reduce the heat to low, remove the lid and cook gently for 20 minutes, or until the onions are tender, golden and syrupy.

2. Stir in the ground spices and cook for 3–4 minutes. Add the tomatoes and aubergine, and cook gently, stirring occasionally, for 6–8 minutes, or until just tender. Keep checking the pan to ensure you don't overcook the aubergine.

3. About 3 minutes before the end of the cooking time, stir in the spinach and cook until it wilts and turns a lovely shade of green. Season with salt and pepper to taste and stir in most of the coriander.

4. Divide the curry between 2 serving plates and sprinkle the remaining coriander over the top.

VARIATIONS

- Instead of ground cinnamon, add a cinnamon stick and remove before serving.
- Add some cardamom pods.
- Add some thickly sliced courgette or cauliflower florets.

Spring Vegetable Broth

This lovely fresh-tasting soup is packed with seasonal spring vegetables and herbs. It's cleansing as well as nutritious – onion and garlic are wonderful foods for a liver detox. The recipe makes enough for four servings but you can cool any soup you don't eat straight away and store it in a sealed container in the fridge for up to three days. Just reheat it in a pan on the hob.

SERVES 4
PREP: 15 MINUTES
COOK: 35 MINUTES

1 onion, diced
1 large leek, washed, trimmed and shredded
2 garlic cloves, crushed
2 celery sticks, diced
300g baby carrots, cut into chunks
1 litre good-quality vegetable stock
1 bay leaf
2 tomatoes*, diced
200g spring greens, shredded
1–3 tsp apple cider vinegar (optional)
a handful of mint, coarsely chopped
sea salt and freshly ground black pepper

1. Put the onion, leek, garlic, celery and carrots in a large saucepan. Add the vegetable stock and the bay leaf and bring to the boil.

2. Reduce the heat to low and simmer gently, stirring occasionally, for 20 minutes, or until the vegetables are tender.

3. Add the tomatoes and spring greens and simmer gently for 10 minutes, or until the vegetables are cooked. Flavour with apple cider vinegar (if using) and stir in the mint. Check the seasoning, adding salt and pepper to taste.

4. Remove and discard the bay leaf and ladle the hot soup into 4 bowls. Enjoy!

VARIATIONS

- Substitute spinach for the spring greens.
- Shallots or spring onions will impart a more delicate flavour than onion.
- Vary the herbs: try chives, parsley, basil, tarragon or dill.

TIP: For a stronger and more distinctive flavour, just before serving drizzle the soup with some oil-free pesto (see page 78).

Really Green Summer Minestrone

Here's a refreshing soup packed with nutritious green vegetables, which are a great source of minerals, including magnesium. You can almost feel it doing you good as it goes down. Ultra-fresh vegetables are natural energisers that revitalise our taste buds too.

SERVES 4
PREP: 15 MINUTES
COOK: 30 MINUTES

4 shallots, peeled and quartered
2 large leeks, washed, trimmed and thinly sliced
2 celery sticks, diced
3 garlic cloves, crushed
900ml good-quality vegetable stock
3 medium courgettes, trimmed and cut into matchsticks
200g asparagus stems, trimmed and cut into short lengths
a handful of rocket, chopped
a bunch of basil, coarsely chopped
1–3 tsp apple cider vinegar
sea salt and freshly ground black pepper

1. Put the shallots, leeks, celery and garlic into a large saucepan. Add the stock and set over a high heat. When it starts to boil, reduce the heat to low.

2. Cover the pan and simmer gently for 15 minutes. Add the courgettes and asparagus and cook very gently for 5–10 minutes, or until all the vegetables are tender.

3. Blitz half the soup with the rocket in a blender or food processor until smooth. Return to the pan and stir well. Add the basil and apple cider vinegar, to taste. Check the seasoning, adding salt and pepper if required.

4. Ladle into 4 serving bowls and serve immediately.

VARIATIONS

- Garnish with rocket leaves or thinly sliced radishes.
- Add some spinach or fresh sorrel leaves or even pak choi.
- Use spring onions instead of shallots.

Winter Vegetable Soup

This soup, made with a selection of seasonal vegetables, will warm you up on a cold day. If you're not going to eat it all immediately, store any leftover portions in sealed containers in the fridge for up to 5 days.

SERVES 4
PREP: 20 MINUTES
COOK: 30–35 MINUTES

1 large onion, finely chopped
2 leeks, washed, trimmed and chopped
2 celery sticks, diced
2.5cm piece of fresh ginger, peeled and diced
2 garlic cloves, crushed
400g carrots, thinly sliced
400g celeriac, cubed
400g butternut squash, peeled, deseeded and cubed
1 litre good-quality vegetable stock
1 bay leaf
a good pinch of dried oregano
200g shredded kale or dark green cabbage
a handful of flat-leaf parsley, chopped
sea salt and freshly ground black pepper

TIP: If you prefer a thicker soup, rather than serving it straight from the pan, you can blitz it in batches in a blender or food processor until smooth. Add some ground spices, if you like.

1. Put the onion, leeks, celery, ginger, garlic, carrots, celeriac and squash into a large saucepan. Add the stock, bay leaf and oregano and set over a high heat.

2. As soon as the soup starts to boil, reduce the heat and simmer gently for 20–25 minutes, or until the vegetables are cooked and tender.

3. Stir in the kale or cabbage and simmer for 5 minutes, or until just tender. Add the parsley and check the seasoning, adding salt and pepper if needed.

4. Ladle the hot soup into 4 serving bowls and serve immediately.

VARIATIONS

- Vary the vegetables: try celeriac, pumpkin, broccoli, cauliflower or spinach.
- Serve with a drizzle of apple cider vinegar.
- Add some sprigs of fresh herbs, e.g. thyme and rosemary.

Smashed Avocado, Tomato & Courgetti Salad

This colourful salad is full of interesting textures and flavours, from the creamy avocado and courgette 'noodles' to the sweet, juicy tomatoes and peppery rocket. It's really quick and easy to make, too! Avocados are a godsend in Phases 1 and 2 when we can't add any oil to our meals, and like olive oil they are rich in monounsaturated healthy fats.

SERVES 2
PREP: 15 MINUTES

2 medium courgettes
1 medium ripe avocado, peeled and stoned
8 cherry or baby plum tomatoes* diced
a small handful of coriander, chopped
baby spinach or rocket leaves, to serve
sea salt and freshly ground black pepper

1. Trim the ends of the courgettes and, using a spiraliser, mandolin or regular potato peeler, spiralise or slice the courgettes lengthways into long thin strips (courgetti). Put the strips into a bowl.

2. In another bowl, mash the avocado with a fork, then add the diced tomatoes and most of the coriander. Season to taste with salt and pepper.

3. Gently toss the courgetti in the smashed avocado and divide the mixture between 2 serving plates. Sprinkle with the remaining coriander and serve immediately with some baby spinach or rocket leaves.

TIP: If the smashed avocado flesh is too thick, you can thin it a little more with some water or apple cider vinegar.

VARIATIONS

- Use parsley, mint or fresh basil instead of coriander.
- Add a crushed garlic clove or some diced spring onion to the smashed avocado.
- You could spiralise carrots instead of courgettes.

Oil-Free Vegetable 'Stir-Fry'

Try substituting stock for oil in all your favourite stir-fries – it works just as well, and seasoned vegetable stock is delicious. For best results, use a non-stick wok or ceramic frying pan. This simple dish proves that you don't need lots of sauces and flavourings to make food taste good – just let the natural flavours of the vegetables shine through. This recipe combines eight different herbs and vegetables (nine if you're using chilli pepper) to keep our friendly gut microbes happy.

SERVES 2
PREP: 15 MINUTES
COOK: 6–7 MINUTES

60–80ml good-quality vegetable stock
4 spring onions, sliced diagonally
2 garlic cloves, crushed
1cm piece of fresh ginger, peeled and diced
a pinch of crushed chilli* or red pepper flakes*
1 small red pepper*, deseeded and cut into strips
250g button mushrooms, halved or quartered
100g thin asparagus spears, trimmed
200g carrots, cut into thin matchsticks
200g small broccoli florets
a small handful of parsley or coriander, chopped
sea salt and freshly ground black pepper

TIP: Taste the stir-fry before adding any salt. There may be enough in the stock to season it.

1. Pour 60ml (4 tablespoons) of the vegetable stock into a large non-stick wok or frying pan and set it over a medium to high heat.

2. When the stock starts to bubble, add the spring onions, garlic and ginger. Cook for 1 minute.

3. Add the chilli or red pepper flakes, pepper, mushrooms, asparagus and carrots and cook for 2 minutes. Add the broccoli florets and, if all the liquid has evaporated in the pan, the remaining stock. Cook for 2–3 more minutes, or until the vegetables are just tender but still have some bite.

4. Check the seasoning, adding black pepper plus some salt, if needed. Sprinkle with the chopped herbs and divide between 2 serving bowls.

VARIATIONS

- Vary the vegetables: try courgettes, cabbage, cauliflower florets, pak choi, green beans, kale or squash.
- Add a large handful of shredded spinach or baby spinach leaves 1 minute before the end of the cooking time. When it wilts and turns bright green, remove the pan and serve.
- Serve the vegetables with courgetti noodles (see page 71).

Spicy Squash & Pumpkin Soup

The pumpkin and butternut squash, and the fresh ginger and turmeric in particular, are good sources of antioxidants and are naturally anti-inflammatory.

SERVES 4
PREP: 20 MINUTES
COOK: 35–40 MINUTES

1 onion, chopped
2 garlic cloves, crushed
2.5cm piece of fresh ginger, peeled and diced
4 large carrots, chopped
3 celery sticks, diced
500g pumpkin or butternut squash, peeled, deseeded and diced
900ml good-quality vegetable stock
2 tsp ground turmeric
½ tsp ground cumin
a good pinch of freshly grated nutmeg
a handful of parsley, finely chopped
sea salt and freshly ground black pepper

1. Put the onion, garlic, ginger, carrot, celery and pumpkin or squash into a large saucepan. Add the stock and set the pan over a high heat. Bring to the boil and then reduce the heat immediately to a simmer.

2. Cook gently for 20–25 minutes, or until all the vegetables are tender.

3. Blitz the soup in batches in a blender or food processor until it is smooth.

4. Return the soup to the pan and stir in the ground spices and nutmeg. Reheat gently, stirring occasionally, over a low heat. Stir in the parsley and check the seasoning.

5. Divide the soup between 4 shallow bowls and serve immediately.

VARIATIONS

- Add some paprika*, cinnamon or ground coriander.
- Use ground ginger instead of fresh.
- Serve topped with sliced mushrooms cooked in stock, or strewn with chopped herbs.

Spaghetti Squash With Kale & Baby Plum Tomatoes

Spaghetti squash is a great veggie substitute for pasta. For best results you need to get the timing just right and take care not to overcook it, so start checking after it's been roasting in the oven for 20 minutes.

SERVES 2
PREP: 20 MINUTES
COOK: 30 MINUTES

1 medium spaghetti squash, split lengthways and deseeded
a pinch of ground cinnamon
100g baby plum tomatoes*, halved
1 red onion, thinly sliced
2 garlic cloves, crushed
a pinch of crushed dried chilli* or red pepper flakes*
75ml good-quality vegetable stock
200g kale, trimmed and coarsely shredded
1–3 tsp apple cider vinegar
a small handful of flat-leaf parsley, chopped, to serve
sea salt and freshly ground black pepper

VARIATIONS

- Use spring greens or spinach instead of kale.
- Add some spiralised carrots or courgettes with the kale.
- Drizzle with some oil-free pesto (see page 78).

1. Preheat the oven to 200°C/180°C fan/gas mark 6. Line 2 baking trays with baking parchment.

2. Prick the skin of the squash a few times with a fork. Dust the inside of each squash half with cinnamon and season lightly with salt and pepper, then place them on one of the lined baking trays, cut-side down. Place the tomatoes, cut-side up, on the other lined baking tray and season with salt and pepper.

3. Bake the squash and tomatoes in the oven for 25–30 minutes, or until the strands of squash are just tender (but not mushy) and the tomatoes are starting to char. Remove from the oven and when the squash is cool enough to handle, scrape out the strands with a fork.

4. Meanwhile, put the onion, garlic, chilli or red pepper flakes and 2 tablespoons of the vegetable stock in a frying pan over a medium heat and cook, stirring occasionally, for 6–8 minutes. Stir in the kale with the remaining stock and simmer gently for 3–4 minutes, or until the kale wilts and the liquid has been absorbed. If it's dry, moisten with another spoonful of stock.

5. Stir in the apple cider vinegar, and fold in the squash strands and tomatoes. Toss together gently.

6. Divide the mixture between 2 serving plates and sprinkle with chopped parsley.

TIP: The key to roasting vegetables without oil is to line the baking trays with baking parchment. Season the vegetables lightly with salt and pepper, then add dried herbs (thyme or oregano) and ground spices (nutmeg, cinnamon, cumin or curry powder).

Breakfast on a Stick

This yummy breakfast is simple to prepare and cooks in a few minutes in a griddle pan or under a hot grill. For HBD purists and anyone who is concerned about nightshades and would like to avoid them, substitute the tomato and peppers for other vegetables on the list.

SERVES 2
PREP: 10 MINUTES
COOK: 6–8 MINUTES

8 medium button mushrooms
6 small cherry tomatoes*
1 green pepper*, deseeded and cut into chunks
1 small red onion, peeled and cut into wedges
1 small courgette, trimmed and cut into chunks
a few sprigs of parsley, chopped
a splash of apple cider vinegar (optional)
sea salt and freshly ground black pepper

1. Preheat the grill to its highest setting.
2. Thread the prepared vegetables and tomatoes on to 2 long or 4 medium kebab skewers, varying the order. Season lightly with salt and pepper.
3. Cook the kebabs under the hot grill for 6–8 minutes, turning them occasionally, until the vegetables are tender and just starting to char.
4. Serve hot, sprinkled with parsley and a splash of apple cider vinegar (if using).

VARIATIONS

- Use button onions, red or yellow peppers*, aubergine* or fennel.
- Sprinkle with some dried herbs or ground spices.
- Add some fresh bay leaves to the skewers.

TIP: You can also cook the kebabs in a non-stick griddle pan over a medium to high heat. Watch them carefully and turn them several times so they don't stick and are cooked through.

NOTE

If you use wooden kebab skewers, soak them in water for 30 minutes before threading them with the vegetables to prevent them burning.

Provençal Vegetable Stew

This summer vegetable stew from the South of France is traditionally served in shallow bowls like a soup and served at room temperature. The fennel, an often-overlooked vegetable, gives this stew a subtle sweetness and succulence. And for the authentic finishing touch, there's an easy to make oil-free pesto here too.

SERVES 4
PREP: 15 MINUTES
COOK: 35 MINUTES

1 onion, finely chopped
3 garlic cloves, crushed
2 celery sticks, finely chopped
1 large leek, chopped
150g fennel bulb, diced
1 green or yellow pepper*, deseeded and diced
300g baby carrots, diced
900ml good-quality vegetable stock
2 courgettes, diced
100g fine green beans, trimmed and diced
sea salt and freshly ground black pepper

OIL-FREE PESTO

1 large bunch fresh basil leaves
3 garlic cloves, peeled
1 tbsp apple cider vinegar
¼ tsp sea salt crystals
1–2 tbsp water

1. Start by making the pesto. In a food processor or blender, blitz the basil, garlic, apple cider vinegar and salt until you have a smooth paste. Add 1–2 tablespoons water to thin it. Transfer to a screw-top jar and keep in the fridge for up to 4 days.

2. To make the stew, put the onion, garlic, celery, leek, fennel, pepper and carrots into a large saucepan with a few tablespoons of the vegetable stock. Set the pan over a low to medium heat and cook, stirring occasionally, for 8–10 minutes until the vegetables are starting to soften. If the stock evaporates quickly, add some more to keep everything moist.

3. Pour in the remaining stock and simmer gently for 15 minutes before adding the courgettes and green beans. Cook for another 5 minutes until all the vegetables are tender. Season to taste with salt and pepper.

4. Ladle the stew into shallow serving bowls and swirl in a spoonful of oil-free pesto.

NOTE

You can freeze the pesto for up to 3 months in ice-cube trays to use as and when needed. The stew will keep well in the fridge in a sealed container for 3–4 days.

VARIATIONS

- Try shredded spring greens or spinach, runner beans or asparagus.
- For colour and sweetness, add 2 diced vine tomatoes*.

TIP: If you don't serve the stew with the oil-free pesto, add chopped fresh basil, mint or flat-leaf parsley before serving instead.

PHASE 2
Reset

Phase 2: Reset

- **14 days of three meals a day**
- **No oil, sugar, grains or alcohol**
- **For best results also avoid dairy and nightshade vegetables**

After two days of vegetables only, day three is a welcome relief, when protein, as well as some fruit, is allowed back in – but remember, you are still oil free.

HUNGER – THE FAT-BURNING FEELING

Reframe any negative feeling of hunger into a positive feeling of, 'This is what burning fat feels like!' It's normal to feel hungry when you start. Sticking to the five-hour rule and combining the right amount of protein and vegetables means very soon you won't feel hungry between meals at all because your blood sugar will be stable and your insulin levels will be lower. And you'll be in the fat-burning zone.

About 50 per cent of HBDers find that if they have yoghurt and fruit for breakfast it doesn't keep them going nearly as well as if they have good-quality protein such as eggs, tofu, fish, meat or chicken. And some find that eating any fruit, other than the daily apple, not only makes them hungrier but also keeps their sweet tooth alive. If you're still hungry a couple of weeks into HBD, try having your apple mid-meal rather than at the end – that can help.

For best results and for pure HBD, avoid all dairy including milk, yoghurt and cheese in Phase 2, even if you don't think you have a problem with it. Dairy is a common allergen and can contribute to digestive complaints (including constipation) as well as skin (eczema, dermatitis), joint or energy issues. Also avoid nightshade vegetables. Potatoes are off the menu anyway, but other nightshades include all kinds of chilli peppers, bell peppers, tomatoes and aubergine.

If you are battling a sweet tooth, avoid yoghurt and fruit for breakfast (dairy avoidance is recommended in any case) and have a good protein breakfast such as eggs or fish (or meat or chicken) or tofu for breakfast. Or have walnuts or seeds with your apple.

ALWAYS WEIGH YOUR FOOD BEFORE COOKING

The weights listed are for raw foods. Remember, variety is key for optimum nutritional benefit – avoid repeating the same protein in your meals. So, if you had chicken for lunch, don't have it again in the evening. The exception is fish, as it's so good for us.

If you weigh less than 65kg:
you can choose to have 120g of protein and 120g of vegetables for each Phase 2 main meal.

If you weigh more than 80kg:
you can have 140g of protein and 140g of vegetables for each Phase 2 main meal.

Breakfast weights remain the same for everyone:
100g of protein and 100g of vegetables.

EGGS

2 eggs for breakfast unless you are under 65kg, in which case you can choose to have 1 egg for breakfast. If you didn't have egg/s for breakfast, you could have 2 eggs for lunch or supper plus your vegetable portion.

CHICKEN AND POULTRY

Avoid the dark meat of poultry, which contains more fat and less protein than breast meat. Also avoid poultry skin, but fish skin, now and again and if you like it, is ok.

MEAT

Cut visible fat off meat. Only eat fresh meat and avoid processed meats including gammon, bacon, Parma ham, bresaola, salami, sausages, etc., until Phase 3 treat meals.

FERMENTED VEGETABLES

These are fine and positively healthy for us, but check the labels if you are eating shop-bought; they should only contain vegetables, salt, maybe herbs, and the live cultures. Homemade is even better.

REMEMBER!

1. **Nothing in between meals except water!** But up to 6 cups of herbal, black or green tea, or coffee can be included with meals. No fruit teas.
2. **Five-hour fast between meals!** Count the 5 hours from when you finish eating, not from when you start.
3. **Something that you fancy is not on the list**. It's not allowed, so avoid it!

When you reach the end of Phase 2 and the end of the 16 first days of the Reset it's time for a celebration. Maybe you didn't think you'd be able to get through, but you did it, you triumphed!

APPLES AND APPLE CIDER VINEGAR

Remember, your apple a day, with any meal, is a must. Apples contain malic acid and so does apple cider vinegar (ACV), and malic acid has both anti-bacterial and anti-fungal properties and may even help to soften gallstones. ACV helps to reduce sugar and alcohol cravings and it's good for digestion too. So eat an apple with one meal every day, and add apple cider vinegar to salads and to water with your meals. Try putting a couple of tablespoons into a wine glass with ice and fizzy water for a festive feeling. But only have it with meals, not in between. There isn't a one-size-fits-all maximum amount of ACV but you'll know you've had too much if you get an acidic feeling in your tummy.

TROUBLESHOOTING in Phase 2

If you come unstuck in Phase 2 and go off the rails, you'll need to start again from the beginning. Treat this phase as sacrosanct. Something magic occurs, a transformation and the beginning of a new you during this time and you need 16 clear days (2 days of Phase 1 and 14 days of Phase 2) to allow the magic to happen and to encourage your body to make the switch from burning carbs for energy to burning fat for energy, and that takes time.

By coming unstuck I don't mean inadvertently eating a food that's not on the Phase 2 list, but any bingeing, drinking alcohol or eating crisps or sugar means starting again.

TRUE HBD

To get the results that you want and deserve, stick 100 per cent with HBD. You're not a true HBDer if:

1. You're in Phase 1 or 2 and doing cardio exercise – cardio can only be reintroduced towards the end of Phase 3.
2. You're skipping breakfast, or any other meal – stick to three meals a day.
3. You're guesstimating rather than weighing out your meals – precision is all-important.
4. You're adding milk (dairy or not) or sugar or sweeteners to tea or coffee – no mini meals.
5. You're drinking tea or coffee between meals – plain water only between meals.
6. You're winging it and copying what others do without having read the book – make HBD your own.

HBDers very often find that when they repeat their annual Reset their weight loss is less dramatic than it was the first time round. Generally it's because we're less toxic and holding on to less water weight than we were when we first started – don't be disheartened!

Quick and easy meals to get you started

BREAKFAST

35g mixed sunflower and pumpkin seeds with 1 grated apple, ½ teaspoon of ground cinnamon, salt and pepper.

••••••

2 soft-boiled eggs with 100g asparagus soldiers.

••••••

75g sugar-free smoked salmon with 50g avocado and 50g salad/raw spinach leaves.

••••••

100g air-fried tofu with 100g grilled or air-fried mushrooms.

LUNCH/DINNER

130g grilled chicken breast with 130g steamed spinach, cabbage and courgette garnished with fresh thyme and/or rosemary.

••••••

160g pre-cooked/canned chickpeas sautéed in stock with herbs, and 130g grilled vegetables, courgette, chicory, radicchio.

••••••

130g sugar-free prawns with 80g avocado and 50g mixed rocket and salad leaves.

••••••

130g spicy mince (chicken, tofu, beef or lamb) sautéed in stock with a mix of 130g veg, e.g. onion, courgette and cabbage with herbs.

PHASE 2 CORE ACTIONS

One type of protein within a meal: no mixed proteins in Phases 2 and 3.

BREAKFAST
100g of one type of protein only per meal and 100g of vegetables.
Or 1–2 eggs with 100g of vegetables.
Or 120g pulses (60g dried weight) with 100g of vegetables.
Aim for at least 3 different vegetables with each meal.

LUNCH/ DINNER
130g of one type of protein per meal (same weight for tofu and tempeh) with 130g mixed veg.
Or 160g pulses (80g dried pulses) with 130g veg.
Aim to include 3 different vegetables or more with each meal.

Permitted Phase 2 Foods	Breakfast quantities	Lunch/dinner quantities
Eggs	2 for breakfast unless you weigh under 65kg, in which case 1 egg	2 eggs for lunch/dinner but not if you had egg/s for breakfast
Cheese – any non-processed, including Cheddar, Gruyère, mozzarella, sheep or goat, feta, halloumi	60g	80g
Yoghurt – full-fat unsweetened, unflavoured sheep, cow, goat, or soya for vegans only *Not to be mixed with other proteins*	160g	Breakfast only
Fish – anchovies (canned in brine or fresh), cod, haddock, halibut, kippers, mackerel, plaice, red mullet, salmon, sardines (canned in brine or fresh), sea bass, sea bream, skate, snapper, sole, tilapia, trout, tuna (canned in water or fresh), turbot	100g	130g
Seafood – clams, crab, lobster, mussels, prawns, shrimps, squid, scallops *Fish and seafood are in separate categories so no mixing within a meal.*	100g	130g
Smoked salmon	75g	100g
Poultry – chicken, turkey, duck breast, white meat only (no skin, legs or wings)	100g	130g
Pork – fresh pork only	100g	130g/Once a week max

Permitted Phase 2 Foods	Breakfast quantities	Lunch/ dinner quantities
Red meat – beef, lamb, venison	100g	130g Twice a week max
Soy – tempeh, tofu	100g	130g
Seeds and nuts – mixed sunflower and pumpkin seeds (ground or whole) or plain walnuts	6 tsp/ 35g	Breakfast only
Pulses – lentils, chickpeas, cannellini, butterbeans, haricot beans, kidney beans	120g soaked/canned or 60g dried	160g soaked/canned or 80g dried
Avocado	up to 80g per meal *If including avocado, weight must be made up by another vegetable in the recipe; extra 20g breakfast*	up to 80g per meal *If including avocado, weight must be made up by another vegetable in the recipe, extra 50g lunch/ dinner*
Black pepper/sea salt		
Apple cider vinegar		
Stock – veg or chicken (homemade/ supermarket liquid only)		
Tamari	HBD purists avoid until Phase 3	HBD purists avoid until Phase 3
Fruit (optional) – blueberries, blackberries, cherries, grapes, mango, papaya, pear, plum, pomegranate, raspberries, strawberries	Up to 100g per meal of one type of fruit as an option	100g

Permitted Phase 2 Foods	Breakfast quantities	Lunch/dinner quantities
Apple – 1 a day in Phase 2, 3 and 4 Lemon – squeezed over fish or chicken (optional) *No mixing fruits within a meal, including berries, and only fruits listed allowed*		1 a day with 1 meal – apple should not be split between 2 meals
Vegetables Artichokes, fresh (hearts bottled in brine, not oil, are ok) Asparagus Aubergine * Bell Peppers* Black coffee Black tea/green tea /herb tea Broccoli Butternut squash Cabbage Capers (sugar-free) Carrots Cauliflower Celeriac Chicory Chillies* Chinese leaves Courgettes Cress Cucumber Endive Fennel Garlic	100g of mixed veg – ideally 3 or more	130g of mixed veg – ideally 3 or more

Permitted Phase 2 Foods	Breakfast quantities	Lunch/dinner quantities
Vegetables continued	100g of mixed veg – ideally 3 or more	130g of mixed veg – ideally 3 or more
Gherkins (sugar-free)		
Ginger		
Green beans		
Herbs – fresh, frozen or dried		
Kale		
Kohlrabi		
Leeks		
Lettuce – all		
Mushrooms		
Okra		
Olives – up to 4 per meal included in veg weight		
Onions		
Pak choi		
Pumpkin		
Radishes		
Rocket		
Romanesco		
Salsify		
Samphire		
Seaweed		
Shallots		
Spinach		
Spring onions		
Sprouts		
Squash		
Tomatoes *	30g max per day	30g max per day

Banned Phase 2 Foods

- No juices
- No alcohol
- No grains
- No sweetcorn
- No sugar, sweeteners, honey
- No oil/fat
- No miso
- No herbs or spices other than fresh, dried or frozen
- No stock cubes, pots or from concentrate – use fresh liquid stock only
- No potato or sweet potato
- No parsnips
- No beetroot
- No fruit tea
- No edamame beans
- No Tabasco

PHASE 2 BREAKFASTS

Green Shakshuka

This traditional Middle Eastern breakfast is a great way to start your day. It's quick, easy and delicious. You can use any green vegetables on the Phase 2 list, provided they total 100g per serving.

SERVES 2
PREP: 10 MINUTES
COOK: 15 MINUTES

60ml good-quality vegetable stock
10g spring onion, thinly sliced
1 garlic clove, crushed
1 red chilli*, diced
75g thin asparagus spears, trimmed and halved
75g broccoli florets
40g spinach, trimmed and chopped
a small handful of flat-leaf parsley or dill
4 organic eggs
sea salt and freshly ground black pepper

1. Set a large non-stick frying pan over a medium heat. Spoon in 1 tablespoon of vegetable stock and 'stir-fry' the spring onion, garlic and chilli, stirring occasionally, for 2–3 minutes or until just tender. If the stock evaporates quickly add a little more to prevent the vegetables sticking to the pan.

2. Meanwhile, blanch the asparagus and broccoli florets in a pan of boiling water for 2 minutes. Drain well.

3. Add the asparagus and broccoli to the frying pan and 'stir-fry' for 2 minutes until they are starting to soften but still slightly crisp, adding more stock when necessary. Stir in the spinach and cook for 2 minutes until it is wilted and bright green, then add the remaining stock. Cook for 2–3 minutes to moisten the mixture, then stir in the herbs. Season to taste with salt and pepper.

4. With the back of a spoon, make 4 hollows in the vegetable mixture and carefully break an egg into each one. Cover the pan and simmer gently over a low heat for 5 minutes, or until the eggs are cooked – the whites should be set and the yolks slightly runny. Serve immediately.

VARIATIONS

- Use crushed dried chilli flakes* instead of fresh chilli.
- Add a pinch of ground cumin, or use different herbs.

Tofu Breakfast Kebabs

These delicious vegan kebabs are easy to assemble and cook on a ridged griddle pan. Tofu is an excellent source of minerals and protein (it's the only vegetarian protein that's comparable to and as good a protein as meat, fish and chicken) and it's filling too. The spicy marinade gives it flavour and colour.

SERVES 2
PREP: 10 MINUTES
COOK: 7–10 MINUTES

200g extra-firm or firm tofu, cubed
2 tsp tamari
100g button mushrooms
60g small cherry tomatoes*
20g small onion or shallot, quartered
paprika*, for dusting
chopped flat-leaf parsley, for sprinkling
a squeeze of lemon juice (optional)
20g rocket
sea salt and freshly ground black pepper

1. Soak 4 thin bamboo or wooden skewers in cold water to prevent them burning on the grill.
2. Put the tofu in a bowl, sprinkle with the tamari and turn the cubes in the marinade. Thread the tofu, mushrooms and cherry tomatoes onto the skewers.
3. Preheat the grill to high, or set a non-stick griddle pan over a medium to high heat.
4. Cook the kebabs for 7–10 minutes, turning occasionally, until the vegetables are just tender and the tofu is golden brown. Dust with paprika, sprinkle with parsley and drizzle with a squeeze of lemon juice (if using). Season with salt and pepper and serve hot with the rocket.

VARIATIONS

- Use courgettes, onion or leek chunks, peppers* or fennel.
- Sprinkle with chopped coriander or snipped chives.
- Use smoked paprika for dusting.

Eggs en Cocotte

This traditional French cooking method using a water bath is a great way to enjoy your breakfast eggs and vegetables. Don't worry if you don't have ramekins – you can use any individual ovenproof dishes or moulds.

SERVES 2
PREP: 5 MINUTES
COOK: 25 MINUTES

150g button mushrooms, chopped
60ml good-quality vegetable stock
50g cherry tomatoes*, chopped
a few sprigs of parsley, chopped
4 organic eggs
sea salt and freshly ground black pepper

1. Preheat the oven to 180°C/160°C fan/gas mark 4.

2. Put the mushrooms and half the vegetable stock into a non-stick frying pan over a medium to high heat. Cook for 8–10 minutes, stirring frequently, until the mushrooms are tender and golden. Add the remaining stock to moisten them if they become dry. Stir in the tomatoes and most of the parsley.

3. Take 2 individual ramekin dishes (large enough to hold 2 eggs each). Divide the mushrooms and tomatoes between them and then carefully crack 2 eggs into each ramekin on top of the mushroom mixture. Sprinkle lightly with salt and pepper and place the ramekins in a roasting pan.

4. Pour boiling water carefully into the roasting pan around the ramekins, until it comes about halfway up their sides.

5. Cook the ramekins in the oven for 15 minutes, or until the whites of the eggs are just set but the yolks are still slightly runny. Sprinkle with the remaining parsley and eat immediately.

VARIATIONS

- Use a sprinkling of dried oregano if you don't have fresh parsley.
- Instead of parsley, use chives, fresh thyme leaves or a few rocket leaves.

Boiled Eggs With Veggie Dippers

A breakfast that's ready in a matter of minutes. Vegetable dippers for soft-boiled eggs make a nutritious alternative to toast. Eggs are not just a good protein food, they're a great source of choline for liver, brain and nervous system health.

SERVES 2
PREP: 5 MINUTES
COOK: 6–7 MINUTES

100g asparagus spears, trimmed
100g fine green beans, topped and tailed
4 organic eggs
sea salt and freshly ground black pepper

TIP: Take care not to overcook the dipping vegetables. They should keep their shape.

1. Blanch the asparagus and green beans in a pan of boiling water for 2–3 minutes, or until they are just tender but still firm (al dente). Remove gently with a slotted spoon, then drain on kitchen paper and season lightly with salt and pepper.

2. Meanwhile, boil the eggs in another pan of water for 4 minutes, or until the whites are set but the yolks are still runny.

3. Place the eggs in egg cups and cut off the tops. Season with a little salt and pepper and serve immediately with the vegetable dippers.

VARIATION

- You can use any vegetables on the Phase 2 food list as dippers. As long as the total weight is 100g per serving, you can use 3 or 4 different vegetables for additional colour and flavour, e.g. cauliflower or broccoli florets, roasted squash, or fennel or carrot sticks.

Breakfast Brochettes

These nutritious chicken or turkey brochettes make a surprisingly filling breakfast to keep you going until lunchtime. If you're always in a hurry first thing in the morning, you could prepare the meat and vegetables the night before and store them in a sealed container in the fridge overnight.

SERVES 2
PREP: 5 MINUTES
COOK: 8–10 MINUTES

200g lean chicken or turkey breast, skinned and cut into chunks
60g small button mushrooms
60g small cherry tomatoes*
80g courgettes, cut into chunks
a squeeze of lemon juice, to serve
snipped chives, for sprinkling
sea salt and freshly ground black pepper

1. Preheat the grill to the highest setting.

2. Thread the chicken or turkey chunks, mushrooms, tomatoes and courgettes on to 2 long or 4 short kebab skewers. Season with salt and pepper.

3. Cook the brochettes under the hot grill for 8–10 minutes, turning occasionally, until the chicken or turkey chunks are slightly browned and cooked right through, and the vegetables are tender. Serve drizzled with lemon juice and a sprinkling of chives.

VARIATIONS

- Vegetarians can substitute tofu for the chicken or turkey meat.
- Vary the vegetables with others on the Phase 2 list, e.g. butternut squash, pumpkin, aubergine* or peppers*.

TIP:
For a slightly chargrilled flavour, cook the brochettes in a non-stick ridged griddle pan. You can moisten them with a little vegetable stock.

TIP:
If you're using wooden skewers, soak them in water first to stop them burning under the grill.

Smoked Salmon With Avocado Salsa

This is a super speedy breakfast – and a fabulous source of good fats – that takes 10 minutes max from prep to table. Smoked salmon is richer and denser than fresh salmon as it contains less water, which is why the serving is 75g instead of 100g. The avocado in the salsa should be just ripe – not too hard or too soft (a tall order, I know); you want the cubes to be tender but to keep their shape.

SERVES 2
PREP: 10 MINUTES

150g thinly sliced smoked salmon
a squeeze of lemon juice
freshly ground black pepper

AVOCADO SALSA
60g red onion, diced
1 garlic clove, crushed
40g ripe tomato*, diced
100g avocado, diced
1 red bird's eye chilli*, deseeded and diced
a few sprigs of coriander, chopped
juice of ½ lemon
sea salt, to taste

TIP: To make this more economical, you can use offcuts instead of the more pricey sliced smoked salmon.

1. Make the avocado salsa: mix the onion, garlic, tomato, avocado, chilli and coriander in a bowl. Stir in the lemon juice and add sea salt to taste.

2. Lay the smoked salmon on 2 serving plates and squeeze some lemon juice over the top. Add a grinding of black pepper and serve with the avocado salsa.

VARIATIONS

- If you don't like coriander, make the salsa with chopped fresh mint or parsley.
- Sprinkle the smoked salmon with a few snipped chives or dill.
- Use spring onions instead of red onions in the salsa.
- Serve with some salad leaves.

Seedy Cinnamon Apple Purée

An apple a day is the only compulsory HBD fruit. Apple pectin fibre has extraordinary detoxifying properties and is also beloved by our friendly gut microbes. If you struggle with eating breakfast (as I do) this slips down easily, I promise, and it keeps you going for ages, too. If you're in a rush you can simply grate the apple and mix it with the seeds and cinnamon – delicious.

SERVES 2
PREP: 5 MINUTES
COOK: 10–15 MINUTES

2 eating apples, cored
2 tbsp water
1 tsp ground cinnamon
70g mixed sunflower and pumpkin seeds

1. Cut the apples into small cubes and place them in a saucepan with the water and cinnamon.

2. Set over a medium heat and cook, stirring occasionally, for 10–15 minutes, or until the apples are tender and can be squashed with a wooden spoon. If they are dry, moisten with a little more water.

3. Remove the pan from the heat and purée the apple with a potato masher, or if you prefer more texture beat with a wooden spoon until it is the right consistency.

4. Set the apple aside to cool a little, then divide between 2 serving bowls. Eat lukewarm or cold, sprinkled with seeds.

TIP: You could make double the quantity of purée and store the extra portions in a sealed container in the fridge for up to 3 days. Or it will freeze well for up to 3 months.

VARIATIONS

- You can also use roasted seeds to add fragrance and crunch.
- Add some grated nutmeg or a pinch of ground ginger.
- Substitute chopped walnuts for the seeds.

PHASE 2
MAINS

TIP: Don't make the kachumber in advance as it will lose its freshness and crunch.

Dhal With Kachumber Salad 

Dhal is the perfect comfort food at the end of a busy day. It's warming, spicy and satisfying and it's delicious with a refreshing Indian kachumber salad. And best of all, you can make the dhal in advance and reheat it. These herbs and spices not only taste delicious, they also provide antioxidants and minerals and are natural anti-inflammatories.

SERVES 2
PREP: 15 MINUTES
COOK: 30–35 MINUTES

80g red onion, diced
2 garlic cloves, crushed
1 tsp grated ginger
1 small red chilli*, diced
300ml good-quality vegetable stock
1 tsp ground turmeric
1 tsp garam masala or chilli powder*
160g split red lentils (dry weight)
juice of ½ lemon
a handful of coriander, chopped
sea salt and freshly ground black pepper

KACHUMBER SALAD
60g tomatoes*, diced
50g cucumber, diced
20g red onion, diced
50g grated carrot
a pinch of ground cumin
a pinch of chilli powder*
2 tsp lemon juice
a few sprigs of mint, chopped
a few sprigs of coriander, chopped

1. To make the dhal, cook the onion, garlic, ginger and chilli in 2–3 tablespoons of the stock in a saucepan over a medium heat, stirring occasionally. Cook for 5 minutes without allowing them to colour.

2. Stir in the ground spices and cook for 1 minute, then add the lentils and remaining stock. Bring to the boil, then reduce the heat and simmer gently for 25 minutes, or until the lentils break up and the dhal thickens – if it's too thick, add some more stock or water. Stir in the lemon juice and season to taste with salt and pepper.

3. While the dhal is cooking, make the kachumber salad: mix all the ingredients together in a bowl, and season to taste with salt and pepper.

4. Serve the hot dhal scattered with chopped coriander and the kachumber salad on the side.

VARIATIONS

- If you like hot and spicy food, add some more chilli.
- You can vary the vegetables in the salad. Try grated radish, cabbage or some blanched green beans.

Butternut Squash & Butterbean Soup

This gently spiced soup is warming but surprisingly filling. The beans supply vegetarian protein and fibre as well as thickening the soup. This quantity serves four people, but you can set aside two or three portions to cool, then store them in individual sealed containers in the fridge for up to four days. Reheat in a pan on the hob.

SERVES 4
PREP: 10 MINUTES
COOK: 35–40 MINUTES

100g onion, finely chopped
2 garlic cloves, crushed
60g carrot, diced
1 litre good-quality hot vegetable stock
360g butternut squash, peeled, deseeded and cubed
1 tsp ground cumin
½ tsp grated nutmeg
½ tsp ground turmeric
640g canned butterbeans, rinsed and drained
a handful of parsley, chopped
sea salt and freshly ground black pepper

1. Put the onion, garlic, carrot and 100ml of the vegetable stock into a large saucepan and bring to the boil. Reduce the heat and simmer for 10 minutes, or until the onion and carrot start to soften and the liquid reduces. Add the squash and cook for 5 minutes, stirring occasionally. Stir in the ground spices and cook for 1 minute.

2. Add the remaining vegetable stock and cook gently for 15–20 minutes, or until all the vegetables are tender. Add half the butterbeans.

3. Blitz the soup in batches in a blender or food processor until it is thick and smooth. Alternatively, use an electric stick blender.

4. Pour the soup back into the pan and stir in the remaining butterbeans. Season to taste with salt and pepper and heat through gently. Stir in most of the parsley.

5. Serve the hot soup sprinkled with the remaining parsley.

VARIATIONS

- You can substitute cannellini or haricot beans for the butterbeans.
- Use pumpkin instead of squash.
- Add some chilli powder or crushed flakes*, ground coriander or cinnamon.
- Instead of parsley, try coriander, dill or mint.

TIP: *Why not measure out a serving of soup, reheat it, then decant it into a thermos flask or insulated bottle and take it to work for a healthy packed lunch?*

Tuscan Bean Soup

This soup is so versatile and you can vary the Phase 2 vegetables according to the season and what's lurking in the fridge! Swap the celery for fennel, for instance, and if you can't find cavolo nero, which is a slightly bitter-tasting Italian cabbage with blackish-green leaves, use any other cabbage or kale, fresh spinach or spring greens instead.

SERVES 4
PREP: 20 MINUTES
COOK: 45 MINUTES

100g onion, diced
50g celery, diced
50g carrot, diced
2 garlic cloves, crushed
1 litre good-quality vegetable stock
120g ripe tomatoes*, chopped
2 sprigs of thyme, leaves stripped
1 sprig of rosemary, leaves stripped and chopped
640g canned cannellini beans, rinsed and drained
200g cavolo nero, washed and shredded
sea salt and freshly ground black pepper

1. Put the onion, celery, carrot and garlic in a large saucepan with 4 tablespoons of the vegetable stock. Set over a low to medium heat and cook, stirring occasionally, for 8–10 minutes, or until the vegetables are softened but not coloured. Add more stock to moisten them if the liquid evaporates too quickly.

2. Add the remaining stock and the tomatoes and bring to the boil. Reduce the heat to a simmer, add the thyme, rosemary and beans and cook gently for 30 minutes, or until all the vegetables are cooked and tender.

3. Add the cavolo nero and simmer for 4–5 minutes – just long enough for it to wilt into the soup without losing its texture and colour. Season to taste with salt and pepper.

4. Ladle the hot soup into 4 serving bowls and serve immediately.

NOTE

You can cool the soup and then divide it up into 4 individual portions. Store them in sealed containers in the fridge for up to 4 days or the freezer for up to 2 months.

VARIATIONS

- Use canned chopped tomatoes* instead of fresh.
- Sprinkle with chopped parsley just before serving.
- Substitute a shredded leek for the onion.

Salmon & Vegetable Parcels

These colourful fish parcels are perfect for a quick and easy supper, and because everything is wrapped in paper or foil and cooked in the oven, there are no fishy smells in the house – and only minimal washing up. Win-win!

SERVES 2
PREP: 15 MINUTES
COOK: 15–20 MINUTES

2 x 130g salmon fillets, skinned
120g carrots, cut into matchsticks
100g courgettes, cut into matchsticks
40g spring onions, thinly sliced
a few chives, snipped
2 garlic cloves, peeled
2.5cm piece of fresh ginger, peeled and cut into matchsticks
2 tbsp good-quality vegetable stock
a few sprigs of coriander, chopped
freshly ground black pepper

1. Preheat the oven to 200°C/180°C fan/ gas mark 6.

2. Cut 2 large squares of baking parchment or kitchen foil and place a salmon fillet in the centre of each one.

3. Scatter the carrots, courgettes, spring onions and chives over the salmon pieces. Tuck in the garlic cloves and ginger, then drizzle with the vegetable stock and season with black pepper.

4. Wrap the paper or foil loosely over the fish and seal the edges securely, twisting them to form 2 neat parcels. Place the parcels on a baking tray and cook in the oven for 15–20 minutes, or until the salmon is cooked through and flakes easily, and the vegetables are just tender.

5. Transfer to 2 serving plates and sprinkle with the chopped coriander. Serve immediately.

VARIATIONS

- Instead of coriander, try flat-leaf parsley or dill.
- You can use any firm-fleshed white fish fillets, including cod, haddock, sole, plaice or sea bass.

Tuna & Vegetable 'Stir-Fry'

Tuna is a good source of protein and very filling. You can use tuna canned in brine if you can't find fresh, and when stir-fried with vegetables in a little stock, it tastes delicious. This quick dish is packed with crunchy vegetables – don't allow them to overcook; they should retain their crispness and 'bite'.

SERVES 2
PREP: 10 MINUTES
CHILL: 10–15 MINUTES
COOK: 6–8 MINUTES

2 x 130g tuna steaks
2 tsp tamari
grated zest and juice of ½ unwaxed lemon
4–5 tbsp good-quality vegetable stock
20g spring onions, thinly sliced
2 garlic cloves, thinly sliced
80g carrots, peeled and cut into thin matchsticks
60g peppers*, sliced
100g tenderstem broccoli, cut into pieces
a few sprigs of coriander, chopped
1 small red chilli*, shredded (optional)
sea salt and freshly ground black pepper

1. Cut the tuna steaks into thin strips and place in a bowl with the tamari and lemon zest and juice. Add a grinding of black pepper. Stir gently to coat the tuna in the marinade, then cover the bowl with cling film and chill in the fridge for 10–15 minutes.

2. Put 3 tablespoons of the vegetable stock in a non-stick wok or frying pan and set over a medium to high heat. Add the spring onions, garlic, carrots, peppers and broccoli and cook briskly for 2–3 minutes. Add more stock if the vegetables look dry and need more moisture.

3. Add the tuna and the marinade to the pan and toss gently with the vegetables. Cook for 3–4 minutes, depending on how well cooked you like your tuna. The vegetables should be slightly tender but still a little crisp. Season with salt and pepper to taste.

4. Divide between 2 serving bowls, sprinkle with coriander and shredded chilli (if using) and serve immediately.

VARIATIONS

- Add some grated fresh ginger for extra aroma.
- If you can't get tenderstem broccoli use small florets from the calabrese variety.

Prawn & Guacamole Salad Wraps

Crisp lettuce leaves are the perfect substitute for grain-based wraps and tortillas. If you've never tried them, now is the time! These wraps are quick and easy to make and taste delicious.

SERVES 2
PREP: 15 MINUTES

260g cooked peeled prawns
50g red or yellow pepper*, deseeded and diced
2 large or 4 medium crisp iceberg lettuce leaves
freshly ground black pepper

GUACAMOLE
1 small green chilli*, diced
10g spring onions, diced
1 garlic clove, crushed
a pinch of sea salt crystals
140g avocado, peeled, stoned and mashed
40g tomato*, deseeded and diced
juice of ½ lemon
a small handful of coriander, chopped

1. Make the guacamole: crush the chilli, spring onions, garlic and salt in a pestle and mortar. Mix with the mashed avocado, tomato, lemon juice and coriander.

2. In a bowl, gently mix the prawns with the red or yellow pepper and the guacamole. Season to taste with black pepper.

3. Spread out the iceberg lettuce leaves on a board and divide the prawn and guacamole mixture between them. Fold the sides of each leaf into the middle to cover the mixture and then fold the top and bottom of each leaf over to make a filled parcel.

4. Wrap some foil or baking parchment around the parcels to make them easier to hold and serve immediately.

VARIATIONS

- Substitute 260g cooked diced chicken breast fillet for the prawns.
- Instead of diced pepper, use raw carrot or cucumber matchsticks.
- Any large crisp lettuce leaves work well – try Cos or even Little Gem.
- Instead of guacamole use avocado salsa (see page 106).
- If you're not a fan of coriander, use flat-leaf parsley instead.

Quick & Easy Aegean Fish Stew

This colourful fish stew makes a surprisingly speedy supper. You can use virtually any firm-fleshed fish fillets. Some fishmongers and fresh fish counters in supermarkets sell ready prepared and mixed fish for soups and stews; remember not to mix fish and shellfish, though, until you reach Phase 4. It's always a good idea to check the fish over for bones before cooking.

SERVES 2
PREP: 15 MINUTES
COOK: 25–30 MINUTES

350ml good-quality fish or vegetable stock
50g onion, diced
160g butternut squash, peeled and diced
1 garlic clove, crushed
50g courgette, sliced
a pinch of crushed chilli flakes* (optional)
a pinch of saffron threads
1 bay leaf
260g mixed firm fish fillets, e.g. cod, sea bream, sea bass, red mullet, salmon, cut into chunks
a handful of flat-leaf parsley, chopped
sea salt and freshly ground black pepper

TIP: The Greeks add tomatoes to the stew, so you could substitute a chopped tomato for the courgette.*

1. Put 3 tablespoons of the stock into a large saucepan over a low to medium heat. Add the onion, butternut squash and garlic and cook for 8–10 minutes, stirring occasionally, until they have softened. If they start to become dry, add a little more stock. Take care that they do not brown.

2. Stir in the courgette, chilli flakes (if using) and saffron. Cook for 1 minute and then add the remaining stock and bay leaf.

3. Bring to the boil, then reduce the heat immediately to barely a simmer and cover the pan. Cook gently for 10 minutes, or until the vegetables are tender. Add the fish and simmer for a further 5 minutes, or until it is just cooked and opaque. Check the seasoning, adding salt and pepper to taste, and stir in the parsley.

4. Ladle into 2 shallow serving bowls, dividing the fish equally between them. Serve immediately.

VARIATIONS

- Use powdered saffron instead of threads.
- Instead of parsley, use chopped dill.
- Substitute mixed seafood (or frozen fruits de mer, defrosted) for the fish. Choose from mussels, prawns, squid and scallops.

Stir-Fried Ginger Chicken & Greens

Stir-fries are so quick and easy to make when you're in a hurry. Using vegetable stock instead of oil works so well and adds a delicious flavour, as well as extra minerals, to the dish. Try it, we think you'll like it!

SERVES 2
PREP: 15 MINUTES
MARINATE: 10 MINUTES
COOK: 10–12 MINUTES

260g chicken breast fillets, skinned and cut into small cubes
2.5cm piece of fresh ginger, peeled and diced or shredded
2 garlic cloves, crushed
1 lemongrass stalk, peeled and thinly sliced
1 tbsp lemon juice
4–5 tbsp good-quality vegetable stock
50g leeks, washed, trimmed and shredded
85g asparagus, trimmed and cut into short lengths
25g celery, cut into short lengths
100g spring greens or spinach leaves, washed and shredded
sea salt and freshly ground black pepper

1. Put the chicken cubes in a bowl with the ginger, garlic, lemongrass and lemon juice. Stir gently to coat the chicken pieces, then set aside in a cool place for 10 minutes to marinate.

2. Put 2 tablespoons of the vegetable stock into a non-stick wok or frying pan and set over a medium to high heat. Add the chicken and 'stir-fry' for 4–5 minutes until cooked through and golden brown all over.

3. Add the leeks, asparagus, celery and greens and 'stir-fry' for 4–5 minutes, adding more stock as necessary to keep the vegetables moist, until they are cooked and the asparagus and celery are just tender but still have some bite.

4. Season to taste with salt and pepper and transfer to 2 serving bowls. Serve immediately.

VARIATIONS

- Use pak choi, Chinese leaves or cabbage instead of spring greens or spinach.
- Vary the other vegetables: try broccoli florets, courgettes, fennel or mushrooms.
- Use turkey breast instead of chicken fillets.

Chicken Paprika

In this yummy take on the classic chicken paprika we're using spiralised vegetable courgetti instead of noodles. They're so healthy and fresh-tasting and add a lovely shade of green (remember that anything green supplies us with energy-giving magnesium).

SERVES 2
PREP: 10 MINUTES
COOK: 40–50 MINUTES

40g onion, chopped
2 garlic cloves, crushed
40g red or green pepper*, deseeded and sliced
180ml good-quality hot chicken stock
2 x 130g chicken breasts, skinned and cubed
1 tsp paprika*
1 tsp caraway seeds
40g chopped tomatoes*
1 bay leaf
a few sprigs of thyme
140g courgettes
a few sprigs of parsley, chopped
sea salt and freshly ground black pepper

1. Put the onion, garlic, pepper and 80ml of the stock into a saucepan. Bring to the boil, reduce the heat to low and cook for 10–15 minutes, stirring occasionally, until the vegetables are tender and most of the liquid has been absorbed.

2. Add the chicken cubes and cook for 5 minutes on all sides until lightly browned. Stir in the paprika and caraway seeds and cook for 1 minute. Add the remaining stock, tomatoes, bay leaf and thyme and bring to the boil.

3. Reduce the heat and simmer gently for 25–30 minutes, or until the sauce reduces and the chicken is thoroughly cooked. Discard the bay leaf and thyme and season to taste with salt and pepper.

4. Meanwhile, make the courgetti. To spiralise the courgettes lengthways, use blade C of a spiraliser or a julienne peeler. You can use a mandolin slicer but you'll have thicker noodles. You can also use an ordinary vegetable peeler. Steam the courgette strips in a steamer or colander over a pan of boiling water for 3–4 minutes until just tender.

5. Divide the paprika chicken between 2 serving plates and sprinkle with the parsley. Serve immediately with the courgetti.

VARIATIONS

- Use turkey breast fillets instead of chicken.
- Use carrots or butternut squash instead of courgettes – adjust the cooking times accordingly.
- Sprinkle with dill or chives instead of parsley.

NOTE

Add more stock if the chicken mixture looks dry.

Warm Lentil Salad

This rustic lentil and squashed tomato salad is bursting with lovely earthy flavours. You can enjoy it at any time of the year. It's best eaten lukewarm or at room temperature.

SERVES 2
PREP: 15 MINUTES
COOK: 30 MINUTES

160g brown or green lentils (dried weight)
80–100ml good-quality vegetable stock
120g red onion, diced
3 garlic cloves, crushed
50g carrot, diced
30g celery, diced
a pinch of crushed red chilli flakes*
60g small cherry tomatoes*
juice of ½ lemon
a handful of basil leaves, torn, plus small sprigs to garnish
sea salt and freshly ground black pepper

1. Put the lentils in a sieve and pick them over to remove any small stones. Rinse under running cold water.

2. Half-fill a saucepan with water and bring it to the boil. Add the lentils, then reduce the heat and simmer for 20 minutes, or until they are just tender but not mushy. Drain in a colander, reserving a little of the cooking water.

3. While the lentils are cooking, put half the vegetable stock into a large non-stick frying pan with the onion, garlic, carrot and celery. Set over a medium to high heat and bring to the boil, then reduce the heat and simmer gently, stirring occasionally, for 15 minutes, or until the vegetables are tender. If the mixture is dry, moisten it with more stock.

4. Add the chilli flakes and whole tomatoes and cook for 5 minutes, or until the tomatoes start to soften. Push down on them with a wooden spoon so they burst. Stir in the drained cooked lentils and cook gently for 10 minutes, or until everything is warmed through and tender. Add about 1–2 tablespoons of the reserved lentil cooking water to make it creamy.

5. Check the seasoning and add salt and pepper to taste. Drizzle with lemon juice and stir in the torn basil leaves. Set aside to cool a little and then divide between 2 shallow serving bowls or plates and scatter the basil sprigs over the top.

VARIATIONS

- Use flat-leaf parsley instead of basil.
- Substitute a leek for the onion.
- Add a pinch of ground cumin or crushed fennel or cumin seeds.

TIP: *Instead of rosemary stems you can use wooden skewers that you have soaked in water for 30 minutes to prevent them burning under the grill.*

Lemon Chicken Souvlaki

This classic Greek dish is easy to make and tastes so good. The spring greens, known as *horta* in Greece, are sometimes served cold and often include wild herbs and dandelion leaves. If you're not a fan of grilled onion, just substitute 100g salad or cooked vegetables from the Phase 2 list.

SERVES 2
PREP: 10 MINUTES
CHILL: 10 MINUTES
COOK: 10 MINUTES

1 tsp fennel seeds
a good pinch of dried oregano
grated zest and juice of 1 unwaxed lemon
260g chicken breast fillets, skinned and cut into chunks
4 woody sprigs of rosemary
100g (1 small) red onion, peeled and cut into chunks
160g spring greens, shredded
2 garlic cloves, crushed
a handful of flat-leaf parsley, chopped
sea salt and freshly ground black pepper

1. Crush the fennel seeds in a pestle and mortar, then transfer them to a bowl. Stir in the oregano, lemon zest and juice. Add the chicken pieces and turn them in the lemony mixture.

2. Strip the leaves from the rosemary stems, just leaving a few on the tips. Thread the chicken pieces and red onion chunks onto the rosemary skewers. Cover and chill in the fridge for 10 minutes.

3. Place the skewers on a lined grill pan and cook under a preheated grill for 10 minutes, turning occasionally, until the chicken pieces are golden brown and cooked right through, and the onion is tender.

4. Meanwhile, steam the spring greens in a steamer basket or colander placed over a pan of boiling water for 2–3 minutes until just tender (they should retain a little bite) and stir in the garlic. Season to taste with salt and pepper

5. Serve the chicken skewers sprinkled with chopped parsley, alongside the steamed spring greens.

VARIATIONS

- Use turkey breast instead of chicken.
- Use kale, fresh spinach or green cabbage instead of spring greens.
- Oregano is traditional but you can substitute thyme.
- Add some fresh bay leaves to the skewers for a more aromatic flavour.

Beefburgers With Portobello Mushrooms

If you like red meat, nothing tastes as good as a homemade beef burger. Roasted Portobello mushroom 'buns' are a great alternative to bread; mushrooms (all mushrooms, even the humble buttons) are a powerful immune tonic. Buy organic when you possibly can.

SERVES 2
PREP: 15 MINUTES
CHILL: 30 MINUTES
COOK: 18 MINUTES

260g very lean (ideally 5–7% fat) minced beef
60g onion, grated
2 garlic cloves, crushed
a few sprigs of parsley, chopped
a pinch of English mustard powder
1 sugar-free gherkin, diced
2 x 80g Portobello mushrooms, wiped
1–2 tbsp vegetable stock
a squeeze of lemon juice
20g rocket leaves
2 cherry tomatoes*, thinly sliced
sea salt and freshly ground black pepper

VARIATIONS

- Use baby salad leaves instead of rocket.
- Use griddled thick slices of aubergine* instead of mushrooms.
- Use large field mushrooms instead of Portobello ones.

1. Mix together the minced beef, onion, garlic, parsley, mustard powder and gherkin in a bowl. Season with salt and pepper. Cover the bowl and chill in the fridge for at least 30 minutes.

2. Meanwhile, preheat the oven to 230°C/210°C fan/gas mark 8.

3. Trim the stems of the mushrooms to allow them to sit flat, gill-side down, on a baking tray lined with baking parchment. Bake in the oven for 8–10 minutes, then turn them over and bake for 8 more minutes until they are tender and juicy.

4. While the mushrooms are cooking, divide the beef mixture into 2 portions with damp hands, and shape each one into a burger.

5. Preheat a non-stick griddle pan or overhead grill. Put the burgers in the hot pan or place them on a rack under the grill and cook for 4–5 minutes on each side until they are cooked through and lightly browned. If they start to stick to the griddle pan, moisten it with a spoonful of vegetable stock.

6. Place the grilled mushrooms, gill-side up, on 2 serving plates. Season with salt and pepper and sprinkle with lemon juice. Add the rocket leaves and sliced tomatoes, then top with the beefburgers and serve immediately.

TIP: If the Portobello mushrooms are not very thick, you could use 4: 2 underneath the burgers and 2 on top like a bun.

Quick Salmon & Broccoli Courgetti

Spiralised vegetables are a positively healthy (and fun) alternative to pasta. We've used courgettes for this simple lunch or supper but butternut squash and/or carrots work equally well.

SERVES 2
PREP: 10 MINUTES
MARINATE: 10 MINUTES
COOK: 10 MINUTES

2 x 130g salmon fillets, skinned
2 tsp lemon juice
4–5 tbsp good-quality vegetable stock
1 small red chilli*, deseeded and diced
60g thin sprouting broccoli, trimmed
50g thin asparagus, trimmed
grated zest and juice of 1 small unwaxed lemon
150g courgettes, cut into long thin strips (see page 124)
sea salt and freshly ground black pepper

VARIATIONS

- Add a little grated fresh ginger.
- Use broccoli florets or tenderstem broccoli instead of sprouting broccoli.
- Substitute thin green beans or pak choi for the asparagus.
- Sprinkle with chopped herbs, e.g. parsley, dill or basil.

1. Put the salmon fillets in a bowl and turn them in the lemon juice. Leave in the fridge or a cool place to marinate for 10 minutes.

2. Set a non-stick frying pan or griddle pan over a medium heat. When it's hot, add the salmon pieces (reserving the marinade) and 2 tablespoons of the vegetable stock. Cook for 2–3 minutes on each side until the salmon is cooked right through. Pour any leftover marinade into the pan and let it bubble for 1 minute. Remove from the heat.

3. Meanwhile, set a non-stick wok or a deep frying pan over a high heat. Add the chilli and 1 tablespoon of stock and cook for a few seconds. Add the broccoli, asparagus and the remaining stock and 'stir-fry' briskly for 3–4 minutes until the vegetables are just tender but still firm with a little bite. Stir in the lemon zest and juice and season to taste with salt and pepper.

4. While the salmon and vegetables are cooking, steam the courgette strips in a steamer or a colander over a pan of boiling water for 3–4 minutes until just tender.

5. Put a salmon fillet on each plate and pour the pan juices over the top. Serve immediately with the courgette strips and other vegetables.

TIP: If the salmon starts to dry out in the pan, add more vegetable stock to moisten it.

PHASE 3
Burn

Phase 3: Burn

- **10 weeks minimum**
- **Extra virgin olive oil is reintroduced**
- **Weekly treat meal**

In Phase 3 we're doing exactly what we've been doing in Phase 2 but with three important changes:

1. Extra virgin olive oil – adding in the oil with superfood status. 1 tablespoon (15–20ml) with each meal.

2. The weekly treat meal – fun and feasting with friends/family and lots of extra calories.

3. Experimenting with foods that were eliminated in Phase 2.

Phase 3 is when life starts to feel more normal again. In Phase 2 we're feeding our body exactly what it needs – real food to reduce inflammation, kickstart weight loss and for repair. But Phase 3 is the time for experimentation – to slowly reintroduce the foods you've avoided up until now and see how your body reacts to them.

EXPERIMENTATION TIME

You might start by introducing yoghurt for breakfast or cheese for lunch. Over the following three days make a note of any changes you feel. Are you bloated? Do you feel any joint or other pain? Nausea? A drop in energy levels? A shift in mood? This is the best way, to listen to feedback from your body, to determine if a food is less than good for you. If you react negatively don't try that particular food again for at least a month. As long as you don't have an autoimmune condition you could try nightshades next – again, giving your body three days to register any changes. Then maybe try wheat at your treat meal. Tune in – how do you feel? How is your energy, mood, bloating?

WEIGHT-LOSS STALL

Weight loss is generally speedy in the 16-day Reset, but it's normal for it to ebb and flow. It often grinds to a halt in the first week or two of Phase 3 and that is completely normal; weight loss is affected by hydration, salt and water retention, hormones, sleep quality and stress. So if the weight stops coming off, check in with yourself and ask:

1. Am I drinking enough water, and am I having most of it before lunch? Remember that dehydration slows our metabolic rate and that morning water is key for good hydration. Also be aware that when we're dehydrated we retain water.

2. Am I getting at least seven hours' sleep? Remember that fat burning happens while we sleep.

3. Am I eating good-quality protein and veg for breakfast? Remember that morning protein helps to reset our leptin sensitivity and stokes our metabolic fire.

4. Am I pooping properly every day (proper poos, not rabbit droppings). Remember that pooping is a vital part of the detox process and if you get backed up your body stops burning fat.

5. Am I focusing on eating mainly green, above-ground vegetables? Remember that root vegetables contain more carbs and therefore more sugar than greens, and more carbs slow down fat burning.

6. Am I getting too much sugar in the form of fruit? If weight loss is your priority have your one apple a day and avoid extra fruit.

7. Am I having a really good treat meal once a week? Remember that if we're taking in fewer calories than we need for too long, the metabolic rate will drop and fat burning will slow or stop.

RYE BREAD

In Phase 3 try 100 per cent rye bread, perhaps with eggs, spinach and avocado, for your weekend breakfast. Again, tune in for feedback from your body. If you find rye suits you, you can have up to 100g in addition to your protein and veg up to twice a week. However, see it as a treat rather than letting it become part of normal life, and remember that if weight loss is your primary concern, rye will slow you down.

HOLIDAYS

What to do if you've had too much fun on holiday and put some weight on? You'll probably find that most of it is water weight and quickly comes off again. A couple of weeks of strict Phase 3 will sort you out – measuring out your olive oil, weighing portions again but still having the all-important treat meal. There's no need to go back to the beginning, just get back to strict Phase 3 and all will be well. But while on holiday it's worth trying to avoid the bread basket and no drinks at lunchtime. HBDers may well go wild on holiday and believe it or not they can't wait to get back to the safety of the HBD rules, and they know that within a few days they will have shed the water excess weight (which is undoubtedly gained from sugary cocktails and snacks).

BEWARE THE SUGAR MONSTER

If you thought you could never escape the clutches of the sugar monster and you got safely through the first 16 days, huge congratulations to you! But my advice to you is to avoid sugar altogether for a while longer. The sugar monster is dormant, not dead, and ready to spring back into action with the first taste. So at treat meals have cheese instead of pudding/dessert. Or nuts with 100 per cent chocolate (no sugar – check the label).

You may now reintroduce dairy (yoghurt and cheese) and nightshades if you have excluded them in Phase 2. Remember to reintroduce one at a time and leave 3 days between each new introduction. Keep a food diary to note any effects, e.g. bloating, joint pain, skin breakouts, blocked sinuses, nausea, fatigue etc.

TREAT MEALS

Here are three reasons why a weekly treat meal is a must in Phase 3 and onwards:

1. If you keep up the low-calorie regime beyond the first two phases your body will adapt by lowering your metabolic rate – fat burning slows down or stops.

2. The effect of giving our body a little weekly shock of 'poison' shakes everything up and restimulates fat burning. This is known as a hormetic effect. Calorie restriction and icy swims also have a hormetic effect in our body – they stimulate the immune system and make us feel perkier and more alive – so enjoy your treat meal and the knowledge that it's doing you good.

3. Feasting with family and friends and laughing and sharing stories at the hearth has always been an important part of human life.

You can have your first treat meal on day 17 if you'd like to, but mostly people hold out until the weekend and regularly have their treat meal on Friday or Saturday night. But equally you could have yours for lunch, or even breakfast. Remember, though, that it's a treat meal rather than a treat day!

Resist the temptation to use carbs and sugar to increase your treat-meal calories, focus on fats. Eating lots of carbs at treat meals results in water retention that can take days to expel. Fats are not the enemies they are made out to be. We need fats and protein for life itself but, believe it or not, we don't need carbs.

But that doesn't mean no treat-meal carbs – think potatoes roasted in goose fat, or chips/fries with mayo (or bearnaise or Hollandaise sauces) or mashed or baked potatoes dripping in butter; these are delicious and necessary high-calorie treat-meal accompaniments. The glut of extra calories once a week in Phase 3 keeps us in fat-burning mode. How about a traditional Sunday roast or an aromatic creamy curry? Or fish and chips (try a gluten-free version for your first treat meal) with mushy peas? Add a cocktail or a couple of glasses of delicious wine, too.

TREAT MEAL CALORIES

A rough rule of thumb is to aim for a minimum of 500 extra calories for this meal, and in this case, more really is better! A treat meal doesn't need to be a junk meal but ideally it would be a high-fat meal; that's the healthiest and easiest way to get your extra calories.

To give you an idea of how to get the extra calories: 30g of walnuts or pine nuts = 200 calories; 50g butter = 350 calories; 100g double cream = 250 calories; 100g camembert cheese = 300 calories; 100g French fries/chips = 300 calories; 200g sirloin steak = 350 calories; 30g hollandaise, bearnaise sauce or mayonnaise = 200 calories; 100g dark chocolate = 500 calories (all these values are approximate).

EATING OUT

You can even eat out in restaurants again and not count these meals as treat meals. Ask for plain, grilled protein foods with salad or green vegetables. You don't want to be weighing out your foods forever, so get your eye used to estimating the weight of your portions at home and then weighing them (remember to weigh your food raw). You'll find you get better and better at this and that will help you to judge restaurant portions too.

EXTRA VIRGIN OLIVE OIL

The oil with superfood status is added back in. There's nothing wrong with ghee or avocado oil or coconut oil in Phase 3 treat meals and occasionally in Phase 4. But for maximum HBD health benefits stick to extra virgin olive oil and use the best quality you can find. Extra virgin olive oil (EVOO) is packed full of plant antioxidants and polyphenols, which are beloved by our friendly gut microbes. It's also anti-inflammatory and associated with a lower BMI, and with cardiovascular health. The best EVOO (don't accept anything less), the highest in antioxidants, is peppery rather than fruity and comes in dark-glass bottles. Add 1 tablespoon (15–20ml) to each meal in Phase 3 (you can have more in Phase 4). There's no need to add it to your breakfasts if you are having yoghurt and fruit or seeds and apple, but the decadence of gently frying your mushrooms and other veg in olive oil is a pleasure indeed.

We somehow got hoodwinked into believing that cooking with olive oil was dangerous, that it turned toxic with heat. And we heard that cooking with extra virgin olive oil was even worse. Turns out that's not true at all! Not only is olive oil one of the safest oils to cook with but extra virgin olive oil contains high levels of antioxidants, which actually protect it from any heat damage.

This makes sense when you think about it – if cooking with olive oil was unhealthy, then the Mediterranean diet would be unhealthy too. What self-respecting nonna would use rapeseed oil or sunflower oil in place of the glorious superfood that's extra virgin olive oil?

EXERCISE

Keep it light until you're at, or very close to, your target weight. Remember that cardio and HBD (until you're in Phase 4) are incompatible. If you're walking or swimming and you can still speak, and you're not breathless, you're in the right zone. Otherwise the exercise is likely to be releasing too much of the stress hormone cortisol. Cortisol not only takes us out of fat-burning mode but it also means we're back in the sympathetic nervous state of flight or fight. And that's absolutely not what we're after in HBD.

REACHED YOUR GOAL?

Once you reach the end of Phase 3 and if you still want to lose more weight just keep going; stay in Phase 3 until you're happy and then move into the Forever Phase.

WHAT DO YOU DO IF YOU GET UNSTUCK/ DERAILED?

There's no need to start from the beginning again (which you would need to do if you broke the rules during the first 16 days). Just take yourself back in hand and revisit your reasons – why you started HBD in the first place. Dust yourself down and get back onto strict Phase 3. It's the same advice for holidays or the occasional wild weekend – don't restart, just get back to Phase 3.

By the time you get to the end of Phase 3 and as long as you've been tuning in to feedback from your body, you will have a clear idea of which foods are good for you and which are best avoided – you've been using your body as a human lab and that will stand you in good stead forever!

TROUBLESHOOTING Phase 3

If you reach your weight-loss goal in less than 10 weeks, don't stop! Stick to the HBD principles but increase your quantities (add 10–20g of protein and the same in veg to each meal) and increase olive oil to 25–30ml with meals. You can also add a small handful of walnuts (no other nuts, because walnuts are higher in anti-inflammatory omega 3 fats) or a tablespoon of sunflower or pumpkin seeds to your meals to increase good fats and calories. A side point – if including walnuts, the inclusion of black pepper improves absorption, and in fact improves nutrient absorption with any meal – so apply liberally! Also increase carby root veg such as beetroot and sweet potatoes. But ONLY do this if you have reached your weight-loss goal.

You might hear people on Instagram talking about Phase 2 with oil and what they mean is Strict Phase 3, which is essentially an extension of Phase 2 and the same portion sizes, but with the addition of olive oil and a weekly treat meal.

PHASE 3 CORE ACTIONS

One type of protein within a meal – no mixed proteins in Phases 2 and 3

Permitted food as in Phase 1 and Phase 2 plus those listed below – weights are the same as for Phase 2

Permitted Phase 3 Foods

- 1 tablespoon extra virgin olive oil per meal (omit this at a yoghurt and fruit breakfast) – compulsory
- 1 weekly treat meal – compulsory
- Skin on chicken/poultry can now be eaten and so can dark meat (legs, wings) – optional
- Meat no longer has to be low-fat
- Up to 100g 100% rye (no other grains) bread as an option once or twice a week. This is in addition to the meal and does not replace the protein or the vegetables – optional
- Tabasco
- Miso
- Beetroot
- Sweet potato
- Parsnips
- Nightshades: Peppers, Aubergine, Tomatoes, Paprika, Chilli, Cayenne pepper
- 85% dark chocolate, 2 squares max after a meal
- Oats: 30g for breakfast only, however if weight loss is your primary aim, avoid. Soak in water overnight and add 10-15g seeds or walnuts and a grated apple to the mix in the morning
- Processed pork: sausages/ bacon/ ham/ gammon/salami – treat meals only

Banned Phase 3 Foods

- No processed foods – none!
- No alcohol – avoid except at treat meals. But one glass of wine with a non-treat meal, if eating out, max twice a week, is acceptable
- No wheat and grains other than rye – treat meal only
- No sugar and honey – treat meal only
- No nuts, other than walnuts – treat meal only
- No potatoes – treat meal only
- No sweetcorn/peas – treat meal only
- No vinegars other than ACV – treat meal only

PHASE 3 BREAKFASTS

Homemade Baked Beans

Canned beans are certainly convenient but they're often high in sugar, so it's better to make your own. These homemade beans are nutritious, taste fabulous and they're ready in a jiffy – just 15 minutes! If you don't have any butterbeans in the cupboard, not to worry. Any canned white beans will work well.

SERVES 2
PREP: 5 MINUTES
COOK: 12–15 MINUTES

2 tbsp olive oil
100g red onion, diced
1 garlic clove, crushed
40g celery, diced
60g fresh cherry or baby plum tomatoes*, diced
a pinch of smoked paprika*
a large pinch of dried oregano
1 tsp tomato purée*
240g canned butterbeans, rinsed and drained
1–3 tsp apple cider vinegar (optional)
a few sprigs of flat-leaf parsley, finely chopped
sea salt and freshly ground black pepper

1. Heat the olive oil in a saucepan over a low to medium heat. Add the onion, garlic and celery and cook for 6–8 minutes, stirring occasionally, until tender.

2. Stir in the tomatoes, smoked paprika and oregano and cook for 2–3 minutes over a medium heat. Add the tomato purée and butterbeans and warm through gently. Add the apple cider vinegar (if using) to taste and some salt and pepper.

3. Divide the beans between 2 serving plates, sprinkle with parsley and serve immediately.

VARIATIONS

- Use canned chickpeas, haricot or cannellini beans.
- Sprinkle with torn basil leaves.
- Serve with a slice of 100% rye bread, toasted.

Smoked Mackerel Breakfast Salad

If you've never fancied fish for breakfast before, try this salad – it could change your mind forever! Like other oily fish, mackerel is nutrient-dense – a valuable source of protein and of eye-, brain- and heart-friendly omega 3.

SERVES 2
PREP: 10 MINUTES
COOK: 3–4 MINUTES

60g greens, e.g. kale or spinach, washed and trimmed
40g red onion, thinly sliced
40g cucumber, sliced
60g cooked beetroot, diced
200g smoked mackerel fillets
lemon juice, for drizzling

SALAD DRESSING
2 tbsp olive oil
1–2 tsp cider vinegar
juice of ½ lemon
a pinch of mustard powder
sea salt and freshly ground black pepper

1. Make the salad dressing: shake all the ingredients vigorously in a screw-top jar until they are well combined. Season to taste with salt and pepper.

2. Put the kale or spinach in a steamer basket or colander over a saucepan of simmering water. Steam for 3–4 minutes, or until the leaves are just tender. Remove and gently pat them dry with kitchen paper. Set aside until cool enough to handle.

3. Put the red onion, cucumber and beetroot into a bowl. Add the kale or spinach and gently toss everything in the salad dressing.

4. Divide the salad between 2 serving plates. Add the mackerel and drizzle with lemon juice. Serve immediately.

TIP: Use plain boiled beetroot, not the sort you buy in vinegar, which is too acidic and will stain the salad red.

VARIATIONS

- Vary the vegetables: try sliced radishes, spring onions, rocket or grated carrot.
- Add a handful of chopped herbs to the salad dressing, e.g. dill, parsley or chives.

Spanish Vegetable Breakfast Tortilla

This Spanish-inspired tortilla, with lots of colourful vegetables, is a healthy way to kickstart your day. Enjoy it for breakfast or a Phase 4 weekend brunch.

SERVES 2
PREP: 10 MINUTES
COOK: 30 MINUTES

2 tbsp olive oil
80g white onion, thinly sliced
80g red or yellow pepper*, deseeded and diced
2 garlic cloves, crushed
40g courgette, sliced
4 organic eggs
a small handful of flat-leaf parsley, chopped
sea salt and freshly ground black pepper

1. Heat the olive oil in a non-stick frying pan over a medium heat. Cook the onion, pepper and garlic, stirring occasionally, for 6–8 minutes, or until they are just tender. Add the courgette slices and cook for 5 minutes, or until they are golden on both sides and all the vegetables are softened.

2. Meanwhile, beat the eggs in a bowl, stir in the parsley and season with salt and pepper. Pour the mixture into the pan and reduce the heat to low. Cook gently for 15 minutes, or until the tortilla is set and golden brown around the sides and underneath.

3. Pop the pan under a preheated grill for 2–3 minutes until the top of the tortilla is set and golden brown. Serve lukewarm or cold, cut into wedges.

VARIATIONS

- Vary the vegetables: try mushrooms or tomatoes*.
- Add a pinch of sweet or smoked paprika* to spice it up.

TIP: The tortilla will be equally delicious the following day.

Veggie Traybake

Everyone loves a traybake. They're so versatile, easy to prepare and cook, and washing up is kept to a minimum. This is a great option for weekend breakfasts.

SERVES 2
PREP: 15 MINUTES
COOK: 25 MINUTES

50g cherry tomatoes*, halved
50g button mushrooms
50g onion, peeled and cut into chunks
4 sprigs of thyme
2 whole garlic cloves, unpeeled
2 tbsp olive oil
a pinch of cumin seeds
120g halloumi, sliced
50g baby spinach leaves
lemon wedges for squeezing (optional)
sea salt and freshly ground black pepper

1. Preheat the oven to 190°C/170°C fan/gas mark 5.

2. Place the tomatoes, mushrooms and onion chunks in a roasting pan, spreading them out in an even layer. Tuck in the sprigs of thyme and garlic cloves and season with salt and pepper. Drizzle the olive oil over the top.

3. Roast in the oven for 15 minutes, then remove and scatter the cumin seeds and halloumi slices over the top. Return to the oven and cook for a further 10 minutes until the vegetables are tender and the halloumi is golden brown.

4. Squeeze the garlic cloves out of their skins and stir them gently into the cheese and vegetables, along with the spinach leaves. These will wilt slightly but retain a lovely green colour.

5. Divide the mixture between 2 serving plates and serve hot with lemon wedges on the side for squeezing (if using).

VARIATIONS

- Drizzle with a little Tabasco* just before serving.
- Vary the vegetables: try sliced courgette, carrot matchsticks, aubergine* cubes or sweet potato chunks.
- Sprinkle with chopped parsley, snipped chives or some basil leaves.
- Add a pinch of crushed red chilli flakes* for extra spice.
- Replace the halloumi with sliced tofu marinated in 1 tablespoon of tamari sauce for 30 minutes. When the vegetables have been roasting for 15 minutes, arrange the tofu on top and bake for a further 10–15 minutes, or until golden brown.

Spinach & Mushroom Scrambled Eggs

Quick and easy to make, perfectly scrambled eggs should have a soft, creamy texture – they should never be watery or start to catch and brown. The secret is to cook them very slowly, stirring all the time.

SERVES 2
PREP: 10 MINUTES
COOK: 8–10 MINUTES

4 organic eggs
2 tbsp olive oil
120g button mushrooms, diced
80g spinach, washed, trimmed and chopped
sea salt and freshly ground black pepper

1. Break the eggs into a bowl. Add a little salt and pepper and beat them with a fork or a balloon whisk.

2. Heat the olive oil in a non-stick saucepan over a medium heat. Add the mushrooms and cook them for 4–5 minutes, or until they are golden brown and tender, stirring occasionally.

3. Stir in the spinach and cook gently for 1–2 minutes, or until it wilts and turns bright green. Reduce the heat to low.

4. Pour the beaten eggs into the pan and stir them gently into the vegetables with a wooden spoon. Keep stirring, lifting and folding the mixture for 2–3 minutes, or until the egg scrambles and sets. As soon as this happens, turn off the heat.

5. Divide the eggs between 2 serving plates and eat immediately.

VARIATIONS

- Stir some snipped chives or chopped parsley into the beaten eggs.
- Add a pinch of freshly grated nutmeg.

Spring Vegetable Omelette

It's always wonderful when spring finally arrives with its bounty of fresh vegetables after the cold winter months of seasonal cabbage and roots. This easy breakfast omelette is a healthy and delicious way to enjoy them.

SERVES 2
PREP: 15 MINUTES
COOK: 15–20 MINUTES

80g baby carrots, halved

60g thin asparagus stems, chopped

4 organic eggs

a small bunch of chives, snipped

2 tbsp olive oil

60g leek, washed and shredded

1 garlic clove, crushed

sea salt and freshly ground black pepper

VARIATIONS

- Add chopped herbs for extra flavour: try parsley, tarragon, basil, fennel fronds or dill.
- Vary the vegetables: try chopped spinach, spring greens or sorrel, baby asparagus or spring onions.
- Add some small cherry tomatoes*.

1. Cook the baby carrots and asparagus stems in a pan of boiling water for 4–5 minutes, or until just tender. Drain and set aside.

2. Beat the eggs in a bowl with the snipped chives and a little salt and pepper.

3. Heat the olive oil in a small non-stick frying pan or omelette pan over a low to medium heat. Add the leek and garlic and cook gently, stirring occasionally, for 5 minutes, or until they have softened. Add the carrots and asparagus stems to the frying pan and cook for 1 minute.

4. Pour in the beaten egg mixture, tilting the pan so that it runs to the sides and under the vegetables. Cook very gently for 5 minutes, or until the omelette is set and golden brown underneath.

5. Pop the pan under a preheated hot grill for 2–3 minutes until the top is set and golden brown.

6. Slide the omelette out of the pan on to a board or plate. Cut it in half and transfer to 2 serving plates. Eat immediately, while the omelette is piping hot.

Tofu Scramble

If you've never eaten scrambled tofu, now's the time to try it. It's so easy and makes a tasty vegan alternative to scrambled eggs. It's packed with protein, too.

SERVES 2
PREP: 10 MINUTES
COOK: 15 MINUTES

2 tbsp olive oil
100g mushrooms, sliced
50g cherry tomatoes*, quartered
50g baby spinach leaves, torn
200g extra-firm or firm tofu
½ tsp ground turmeric
¼ tsp ground cumin
¼ tsp smoked paprika*
a good pinch of garlic powder
1 tbsp nutritional yeast (for vegans only)
a dash of tamari (optional)
a small handful of parsley, chopped
lemon juice, for drizzling (optional)
sea salt and freshly ground black pepper

1. Heat 1 tablespoon of the olive oil in a frying pan over a medium heat. Cook the mushrooms, turning them occasionally, for 3–4 minutes until golden. Add the tomatoes and spinach and cook for 2 more minutes. Season to taste with salt and pepper.

2. Heat the remaining olive oil in another pan over a medium heat. Use your hands to crumble the tofu into the pan. Stir gently and cook for 4 minutes, or until any water from the tofu has evaporated.

3. Stir in the ground spices, garlic powder and nutritional yeast, if using. Cook for 5 minutes, stirring all the time. At this point, you can add the tamari (if using) for extra flavour.

4. Divide the vegetables between 2 serving plates or shallow bowls. Spoon the tofu scramble over the top and sprinkle with the parsley. Serve immediately, drizzled with lemon juice, if using.

TIP: You can either crumble the tofu with your hands or put the tofu block into the pan and mash it with a potato masher or a fork.

VARIATIONS

- Vary the vegetables: try spring onions, asparagus, courgette or red pepper*.
- Substitute spring greens or curly kale for the spinach. Or add some rocket.
- Experiment with different herbs: try coriander, chives or tarragon.
- Serve with sliced avocado.

PHASE 3
MAINS

Chargrilled Vegetable Halloumi Kebabs With Kohlrabi Slaw

Halloumi cheese is the perfect choice for veggie kebabs because it is very firm and keeps its shape when cooked. If you've never had kohlrabi, you're in for a pleasant surprise. It's an ideal alternative to cabbage or Chinese leaves in slaws – crisp with a slightly peppery flavour. Kohlrabi is a great source of fibre and related to cabbage, broccoli, cauliflower, kale and Brussels sprouts.

SERVES 2
PREP: 20 MINUTES
COOK: 8–10 MINUTES

60g red pepper*, deseeded and cut into chunks
60g button mushrooms
40g courgette, thickly sliced
160g halloumi cheese, cubed
1 tbsp olive oil
a good pinch of dried oregano
sea salt and freshly ground black pepper

KOHLRABI SLAW
100g kohlrabi, trimmed, bulb peeled and thinly sliced into matchsticks
a small handful of mint, chopped
juice of ½ lemon
1 tbsp olive oil
1–2 tsp cider vinegar
1 tsp whole-grain mustard

NOTE

You can also cook the kebabs in a non-stick griddle pan set over a medium heat.

1. Make the kohlrabi slaw: place the kohlrabi sticks in a bowl with the chopped mint. Whisk the other ingredients together until well combined and pour over the kohlrabi and herbs. Toss gently to coat the kolrabi in the dressing and check the seasoning.

2. Thread the pepper pieces, mushrooms, courgette and halloumi onto 2 kebab skewers. Brush the halloumi and vegetables with the olive oil, dust with oregano and season lightly with salt and pepper.

3. Cook the kebabs under a preheated grill set to high, or over hot coals on a barbecue, for 8–10 minutes, turning occasionally, until the vegetables are tender and the halloumi is golden brown.

4. Serve the kebabs immediately with the kohlrabi slaw.

VARIATIONS

- Use chopped parsley and dill instead of mint in the kohlrabi slaw.
- Add a pinch of cumin or fennel seeds to the slaw.
- Add some grated carrot to the slaw.

TIP: If using wooden or bamboo skewers, soak them first in water to prevent them burning.

Spiced Chickpea & Roasted Vegetable Buddha Bowl

Healthy veggie Buddha bowls are all the rage. All you have to do is combine some vegetable protein – such as beans, lentils or tofu – with vegetables and aromatic herbs or spices. Packed with nutritious goodness and flavour, this is food for your soul!

SERVES 2
PREP: 15 MINUTES
COOK: 30 MINUTES

320g canned chickpeas, rinsed and drained
½ tsp smoked paprika*
¼ tsp ground cumin seeds
a pinch of chilli powder*
2 tbsp olive oil
60g red onion, cut into 4 thin wedges
100g pumpkin or butternut squash, peeled and cubed
50g courgette, thickly sliced
2 whole garlic cloves, unpeeled
3 sprigs of rosemary
50g kale, trimmed, large stems removed
a handful of coriander, chopped
apple cider vinegar, for drizzling
sea salt and freshly ground black pepper

VARIATIONS

- Use fresh spinach instead of kale.
- Experiment with sliced aubergine*, fennel and cherry tomatoes*.
- Instead of rosemary, use fresh thyme or oregano sprigs.

1. Preheat the oven to 200°C/180°C fan/gas mark 6.

2. Put the chickpeas in a bowl with the smoked paprika, cumin seeds and chilli powder. Add a pinch of sea salt and toss gently in 1 tablespoon of the olive oil.

3. Spread out the chickpeas in a single layer on a baking tray and roast them in the preheated oven for 15–20 minutes, turning them once or twice, until they are golden brown and slightly crisp. Remove from the oven and set aside to cool.

4. Meanwhile, put the red onion, pumpkin or squash and courgettes in a large roasting pan. Tuck the garlic and rosemary under the vegetables. Drizzle with the remaining olive oil and season with salt and pepper.

5. Roast the vegetables for 25 minutes, or until they are tender. Turn once or twice during cooking to ensure that they cook evenly. Stir in the kale and return to the oven for 5 minutes.

6. Remove the rosemary and squeeze the garlic flesh out of the skins. Stir the garlic and coriander into the roasted vegetables.

7. Divide the spiced chickpeas and roasted vegetables between 2 serving bowls. Drizzle with apple cider vinegar and enjoy!

Jamaican Jerk Prawns With Cauli Rice

Make your own jerk paste for this healthy supper – just blitz everything in a blender. If you use Scotch bonnet chillies the paste will be fiery and authentic, while the herbs and spices such as nutmeg, cinnamon and ginger not only add flavour, they are full of health-giving plant antioxidants, too.

SERVES 2
PREP: 15 MINUTES
CHILL: 30 MINUTES
COOK: 10–12 MINUTES

260g large raw prawns, shelled
1 tbsp olive oil
sea salt and freshly ground black pepper

JERK PASTE
1 tsp allspice berries
1 tsp black peppercorns
1 garlic clove, crushed
2.5cm piece of fresh ginger, peeled and diced
1 hot chilli*, diced (preferably Scotch bonnet)
a good pinch each of ground cinnamon and nutmeg
3 sprigs of thyme, leaves stripped
1 tbsp tamari
juice of 1 small lemon

CAULI 'RICE'
185g cauliflower florets
1 tbsp olive oil
1 lemongrass stalk, peeled and diced
30g spring onions, thinly sliced
45g grated courgette
a handful of coriander, finely chopped
1 tbsp tamari
squeeze of lemon juice

1. Make the jerk paste: crush the allspice berries and peppercorns in a pestle and mortar. Blitz to a paste with the other ingredients in a blender. Transfer the paste to a bowl and add the prawns. Stir gently until they are coated. Cover and chill in the fridge for at least 30 minutes.

2. Next make the cauli rice: pulse the cauliflower florets in a food processor until they have the consistency of small grains of rice. Heat the oil in a frying pan set over a medium to high heat. Stir-fry the lemongrass and spring onions for 2–3 minutes. Add the cauliflower and stir-fry for 5 minutes until it is tender but still slightly crunchy.

3. Transfer to a bowl and mix with the courgette and coriander. Blend the tamari and lemon juice, stir in gently, then cover the bowl with cling film and chill in the fridge for at least 15 minutes.

4. Heat the oil in a frying pan or a griddle set over a medium to high heat. Add the prawns and cook for 1–2 minutes on each side until they turn pink.

5. Serve the jerk prawns immediately with the cauli rice.

VARIATIONS

- Use mint instead of coriander in the cauliflower rice.
- Vary the vegetables: add some chopped spring onions or grated carrot.

TIP: Make sure you do not overcook the prawns – they will lose their juicy tenderness and taste dry and hard. Cook for just 1–2 minutes on each side until they turn pink.

Summer Salmon Parcels

Salmon served with a lovely peppery rocket salad is the perfect dish for a hot summer's day. If wished, you can cook the salmon in advance and serve it cold later. It will keep well in a sealed container in the fridge for 24 hours.

SERVES 2
PREP: 10 MINUTES
COOK: 15–20 MINUTES

2 x 130g salmon fillets, skinned
a few sprigs of tarragon and dill
juice of ½ lemon
100g thin green beans, trimmed
80g baby carrots
sea salt and freshly ground black pepper

ROCKET SALAD
80g wild rocket
2 tbsp olive oil
½ tbsp apple cider vinegar
a pinch of mustard powder
1 small garlic clove, crushed

1. Preheat the oven to 180°C/160°C fan/gas mark 4.

2. Place the salmon fillets on a sheet of kitchen foil with the herbs. Sprinkle with the lemon juice and season lightly with salt and pepper. Fold the foil over the salmon to make a sealed parcel. Place on a baking tray.

2. Cook in the oven for 15–20 minutes until the salmon is cooked through and tender.

3. Meanwhile make the rocket salad: put the rocket in a bowl, then whisk together the olive oil, vinegar, mustard powder and garlic until well combined. Toss the rocket in the dressing.

4. Steam or boil the green beans and baby carrots until they are just tender.

5. Serve the salmon hot, lukewarm or cold with the beans and carrots and rocket salad.

VARIATIONS

- Poach the salmon in some water or stock instead of baking it.
- Serve with any seasonal summer or salad vegetables on the Phase 3 list.
- Use any summer herbs: try dill, chives, mint or chervil

Spinach & Chickpea Curry in a Hurry

This curry is delicately flavoured and quite mild. It's simple and quick to prepare and cook when you return home after a busy day at work. If you want to increase the heat, add a diced fresh chilli* or some dried crushed chilli flakes*.

SERVES 2
PREP: 10 MINUTES
COOK: 20 MINUTES

2 tbsp olive oil
60g red onion, finely chopped
2 garlic cloves, crushed
2cm piece of fresh ginger, peeled and diced
1 tsp mild curry powder
60g canned chopped tomatoes*
100ml good-quality vegetable stock
320g canned chickpeas, rinsed and drained
140g spinach, washed, trimmed and coarsely shredded
a handful of coriander, chopped
sea salt and freshly ground black pepper

1. Heat the olive oil in a saucepan over a low to medium heat. Cook the onion, garlic and ginger, stirring occasionally, for 6–8 minutes, or until softened but not browned. Stir in the curry powder and cook for 1 minute.

2. Add the tomatoes and stock and turn up the heat to medium. Cook for 5 minutes, or until the liquid evaporates and reduces. Stir in the chickpeas and spinach. Cook for 4–5 minutes until the chickpeas are heated through and the spinach wilts. Stir in the coriander and season to taste with salt and pepper.

3. Divide the curry between 2 serving bowls or plates and serve.

VARIATIONS

- Use garam masala and mixed ground spices instead of ready-made curry powder.
- Use kale instead of spinach.
- If you don't like fresh coriander, substitute parsley.
- Use chopped fresh tomatoes* instead of canned.

Thai Courgetti With Salmon Fillets

Spiralised vegetables, such as courgettes or carrots, are fun to eat, taste delicious and make a healthy addition to any meal. Don't worry if you don't have a spiraliser, you can use a julienne peeler, mandolin slicer or even a thin potato peeler.

SERVES 2
PREP: 15 MINUTES
CHILL: 10–15 MINUTES
COOK: 10 MINUTES

100g courgettes
1 tbsp apple cider vinegar
1 tbsp tamari
2 tsp olive oil
2 x 130g salmon fillets
1 red chilli*, deseeded and diced
100g thin green beans, trimmed
60g pak choi, halved lengthways
a few sprigs of coriander, chopped

1. Use a spiraliser to cut the courgettes lengthways with blade C into long thin strands. If you don't have one, slice them lengthways very thinly with a peeler or mandolin slicer.

2. Put the courgette strands in a bowl with the vinegar and tamari and toss them gently to avoid breaking them. Cover the bowl with cling film and chill in the fridge for 10–15 minutes.

3. Heat the oil in a frying pan set over a medium heat. When it's hot, cook the salmon fillets for 3–4 minutes on each side until cooked through. Remove from the pan and cut them into slices. Keep warm.

4. Reduce the heat to barely a simmer and add the courgette strands, chilli and marinade to the pan. Warm through very gently for 1–2 minutes. The courgettes should be slightly tender but still firm.

5. Meanwhile, steam the green beans and pak choi in a steamer basket or colander over a pan of boiling water until the beans are tender and the pak choi is just wilted.

6. Divide the courgetti between 2 serving plates and top with the salmon. Sprinkle with coriander and serve immediately with the green beans and pak choi.

VARIATIONS

- Use white fish fillets instead of salmon.
- Vary the vegetables: try carrots, broccoli, spinach or Romanesco.

White Fish With Ratatouille 

Succulent white fish fillets and aromatic ratatouille flavoured with basil go so well together. I use rolled up delicate plaice or sole fillets but you could also lay skinned cod or haddock fillets on top of the vegetable mixture. If you are cooking this as a special meal, use the ratatouille as a filling, rolling the fish around it.

SERVES 2
PREP: 15 MINUTES
COOK: 20 MINUTES

2 x 130g skinned plaice or lemon sole fillets
juice of ½ lemon
good-quality vegetable stock, if needed
a handful of flat-leaf parsley, chopped

RATATOUILLE
2 tbsp olive oil
40g red pepper*, deseeded and chopped
100g butternut squash, peeled, deseeded and cubed
80g courgette, diced
40g cherry tomatoes*, diced
a few sprigs of basil leaves, chopped or torn
sea salt and freshly ground black pepper

1. Make the ratatouille: heat the oil in a frying pan over a medium heat. Add the pepper, squash and courgette and cook, stirring occasionally, for 8–10 minutes until the vegetables start to soften. Add the tomatoes and cook gently over a low heat for 5 minutes, or until the mixture is starting to reduce and thicken. Stir in the basil and season to taste with salt and pepper.

2. Meanwhile, lay the fish fillets flat on a clean surface and carefully roll them up from the narrow end to the wide end. Secure them with wooden cocktail sticks and place them in the pan on top of the ratatouille. Sprinkle them with lemon juice.

3. Cover the pan with a lid and cook gently for 5 minutes, or until the fish is cooked, looks opaque and flakes easily. Check after 3–4 minutes and if the ratatouille needs more moisture, add a spoonful or two of vegetable stock.

4. Divide the ratatouille between 2 serving plates. Serve topped with the fish rolls, sprinkled with parsley.

VARIATIONS

- Use sea bass fillets instead of plaice or sole.
- Use pumpkin, aubergine* or green or yellow peppers* in the ratatouille.
- Flavour the ratatouille with a few drops of lemon juice.

Pot-Roast Chicken & Spring Vegetables

This simple supper dish is bursting with fresh spring flavours and colours. If you don't have all the vegetables listed, don't worry, you can mix and match using any veggies on the Phase 3 list.

SERVES 2
PREP: 10 MINUTES
COOK: 40–45 MINUTES

2 tbsp olive oil
50g leek, trimmed and sliced
30g celery, sliced
100g baby carrots, whole or halved
2 garlic cloves, crushed
2 x 130g chicken breast fillets, skinned
30g shallots, peeled
300ml good-quality chicken stock
1 bay leaf
a pinch of dried mixed herbs
50g asparagus stems, sliced
a handful of flat-leaf parsley, tarragon, mint or dill, finely chopped
sea salt and freshly ground black pepper

1. Heat the olive oil in a large heavy saucepan or flameproof casserole dish over a low heat. Cook the leek, celery, carrots and garlic, stirring occasionally, for 10 minutes.

2. Increase the heat to medium and add the chicken breasts and shallots. Cook for 5 minutes, turning halfway through, until the chicken pieces are golden brown on both sides and the shallots are starting to colour.

3. Add the stock, bay leaf and mixed herbs, then cover the pan with a lid. Increase the heat and bring to the boil, then reduce the heat to low and simmer gently for 10 minutes. Add the asparagus stems and continue cooking for 15–20 minutes, or until the chicken is cooked through and the vegetables are tender.

4. Discard the bay leaf, stir in the chopped fresh herbs and season to taste with salt and pepper. Divide between 2 serving plates or shallow bowls and serve immediately.

VARIATIONS

- Add some chopped asparagus or green beans.
- Stir in some oil-free pesto (see page 78) just before serving.
- Use turkey breast fillets instead of chicken.
- Use vegetable instead of chicken stock.

Chicken Arrabbiata

Tender chicken cooked in a fiery vegetable sauce is an Italian classic. It's easy to prepare and cook when you get home from work, and ready in just over half an hour.

SERVES 2
PREP: 15 MINUTES
COOK: 25–30 MINUTES

2 x 130g chicken breast fillets, skinned
2 tbsp olive oil
2 garlic cloves, thinly sliced
1 red chilli*, shredded
100g red onion, chopped
60g cherry tomatoes*
2 tsp capers in brine, drained and rinsed
a few sprigs of basil, torn
50g thin green beans, trimmed
50g spinach leaves, washed, trimmed and coarsely chopped
sea salt and freshly ground black pepper

1. Place the chicken breasts between 2 sheets of baking parchment or cling film and flatten them with a meat mallet or rolling pin until they are approximately 1cm thick.

2. Heat the olive oil in a large frying pan over a medium heat. Add the chicken breasts and cook for 3–4 minutes on each side until they are golden brown. Remove from the pan and keep warm.

3. Add the garlic, chilli and onion and cook, stirring occasionally, for 6–8 minutes, or until softened. Stir in the cherry tomatoes and capers and return the chicken to the pan. Reduce the heat to low and simmer gently, stirring occasionally and squashing the tomatoes with a spatula, for 10–15 minutes, or until the chicken is cooked through and the sauce has thickened. Season with salt and pepper and stir in the basil.

4. Meanwhile, cook the green beans in a pan of boiling water for 4 minutes, or until they are tender but still slightly crisp. Drain well.

5. Put the spinach in a colander and pour boiling water over it until it wilts and turns bright green. Press down on it with a saucer to squeeze out the water.

6. Transfer to 2 serving plates and serve immediately with the green beans and spinach.

VARIATIONS

- If you like hot, spicy food, add another chilli or some crushed dried chilli flakes*.
- Serve with any vegetables on the Phase 3 list.
- Substitute flat-leaf parsley for the basil

TIP: Make sure that the burgers are thoroughly cooked – they should not be pink in the middle.

Spicy Chicken Burgers on Aubergine Sliders

Why not substitute roasted or griddled aubergine slices for the usual burger bun? They taste delicious and are light and healthy. Vary the spices in the burgers, depending on how hot or aromatic you like them. And remember, if you're avoiding nightshade veg you can use mushrooms for the bun.

SERVES 2
PREP: 15 MINUTES
COOK: 20 MINUTES

260g chicken breast fillets, skinned and cut into chunks
10g spring onions, trimmed
2 garlic cloves, crushed
¼ tsp ground cumin
½ tsp paprika*
a pinch of crushed chilli flakes* (optional)
a few sprigs of coriander, chopped
4 x 30g slices aubergine* (1cm thick)
2 tbsp olive oil
sea salt and freshly ground black pepper

STIR-FRIED VEGETABLES
1 garlic clove, thinly sliced
50g small broccoli florets
80g mushrooms, thinly sliced
1 tbsp tamari

VARIATIONS

- Use turkey breast instead of chicken.
- Stir-fry any of the following: pak choi, carrots, celery, fennel, ginger, cauliflower, okra or peppers*

1. Preheat the oven to 200°C/180°C fan/gas mark 6. Line a baking tray with foil.

2. Put the chicken chunks, spring onions, garlic, spices and coriander into a food processor or blender and pulse until you have a coarse paste. Season lightly with salt and pepper. Divide the mixture into 2 portions and, using your hands, mould each one into a burger.

3. Place the aubergine slices on the baking tray and brush them lightly with a little of the oil. Season with salt and pepper. Bake them for 7–8 minutes, then turn them over and cook the other side for 7–8 minutes until they are tender and starting to crisp.

4. Meanwhile, lightly brush a non-stick frying pan or griddle pan with a little oil and set it over a medium heat. Cook the burgers for 6–8 minutes on each side, or until they are golden brown on the outside and cooked right through.

5. Make the stir-fried vegetables: heat the remaining oil in a wok or deep frying pan over a medium to high heat. Cook the garlic for 30 seconds, then add the veg and stir-fry for 2–3 minutes until they are just tender but still crisp. Add the tamari sauce.

6. Serve the chicken burgers sandwiched between 2 roasted aubergine sliders with the hot stir-fried vegetables on the side.

Rosemary Chicken & Vegetable Traybake

One of the easiest suppers of all is to cook everything in the oven in one roasting pan. This fragrant mixture of vegetables, fresh herbs and chicken is a great standby. It only take a few minutes to assemble and then you can sit back and relax with your feet up while it cooks.

SERVES 2
PREP: 10 MINUTES
COOK: 30 MINUTES

2 x 130g chicken breast fillets, skinned
60g red onion, cut into wedges
60g red or yellow pepper*, deseeded and cubed
100g courgette, cut into chunks
40g fennel bulb, sliced
a few sprigs of rosemary, plus a few leaves
4 unpeeled garlic cloves
2 tbsp olive oil
a small handful of flat-leaf parsley, chopped
sea salt and freshly ground black pepper

1. Preheat the oven to 200°C/180°C fan/gas mark 6.

2. Slash the top of the chicken breasts a couple of times with a sharp knife. Put them in a large roasting pan with the onion, pepper, courgette and fennel.

3. Tuck the rosemary sprigs and garlic cloves into the gaps between the vegetables and drizzle everything with the olive oil. Season lightly with salt and pepper.

4. Roast in the oven for 30 minutes, or until the chicken is cooked through and golden brown and the vegetables are tender. Discard the rosemary sprigs if they are browned. Squeeze the garlic flesh out of the cloves over the chicken and vegetables. Sprinkle with rosemary leaves and parsley.

5. Divide the chicken and vegetables between 2 serving plates and serve hot or lukewarm.

VARIATIONS

- Drizzle with lemon juice before roasting or just before serving.
- Vary the vegetables: try cherry tomatoes*, aubergine* or butternut squash.
- Use thyme instead of rosemary for a softer, less aromatic flavour.

Griddled Chicken & Sweet Potato Salad Bowl

This colourful salad bowl is delicious served lukewarm or you can eat it cold on a hot summer's day. It also works well for picnics and packed lunches. The lemony dressing is oil-free and can be used for tossing salads and dressing vegetables during Phases 1 and 2.

SERVES 2
PREP: 15 MINUTES
COOK: 25–30 MINUTES

1 tsp coriander seeds
2 tsp cumin seeds
140g yellow or purple sweet potato, peeled and cut into thick matchsticks
60g red onion, cut into small wedges
2 tbsp olive oil
2 x 130g chicken breast fillets, skinned
30g cucumber, thickly sliced
30g baby spinach leaves
a few sprigs of flat-leaf parsley, chopped
sea salt and freshly ground black pepper

LEMONY DRESSING
1 garlic clove, crushed
1 tsp grated fresh ginger
1 tsp mustard
juice of 1 lemon
2 tbsp water

1. Preheat the oven to 200°C/180°C fan/gas mark 6.

2. Make the lemony dressing: blitz all the ingredients in a blender until smooth.

3. Coarsely crush the coriander and cumin seeds in a pestle and mortar. Put the sweet potato and red onion into a roasting pan and sprinkle with the seeds. Drizzle with most of the olive oil and season with salt and pepper. Roast for 25–30 minutes, turning once or twice, until the sweet potato is tender and the onion wedges have softened.

4. Meanwhile, lightly brush a griddle pan with the remaining olive oil and set over a medium heat. Cook the chicken breasts for 6–8 minutes on each side until they are golden brown, attractively striped and cooked right through. Cut them into slices.

5. Put the roasted vegetables, cucumber and spinach in a bowl and toss them lightly in the dressing.

6. Divide the dressed vegetables between 2 shallow serving bowls and top with the sliced chicken. Sprinkle with parsley and serve.

VARIATIONS

- Use butternut squash or pumpkin instead of sweet potato.
- Vary the salad vegetables: try raw courgette sticks, beetroot or rocket.

Italian Chicken Piccata

Serve the chicken with low-carb cauliflower 'rice'. It sounds strange but nothing could be easier – just blitz in a food processor, then heat through gently in a pan. It's the perfect accompaniment for this lemony chicken. Cauliflower is related to cabbage and broccoli and excellent for liver health, whereas rice only supplies carbs.

SERVES 2
PREP: 20 MINUTES
COOK: 15 MINUTES

2 x 130g chicken breast fillets, skinned
2 tbsp olive oil
2 garlic cloves, crushed
2 tbsp capers in brine, drained
grated zest and juice of 1 lemon
1 lemon, sliced
150ml good-quality chicken stock
a small handful of flat-leaf parsley, chopped
60g rocket
sea salt and freshly ground black pepper

CAULIFLOWER RICE
200g cauliflower florets
a pinch of dried crushed chilli flakes*

VARIATIONS

- Add a pinch of paprika* or some freshly chopped herbs to the cauli rice to flavour it.
- You can serve this with salad leaves or vegetables.

1. Slice each chicken breast through the middle so it opens up like a book. Place between 2 sheets of baking parchment or cling film and flatten them with a rolling pin to get 2 thin escalopes.

2. Heat most of the olive oil in a large non-stick frying pan over a medium heat. Add the chicken escalopes to the pan and cook for 4–5 minutes until they are golden brown underneath. Turn them over and cook the other side. Remove from the pan, cover with foil and keep warm.

3. Add the garlic to the pan and cook for 1 minute without browning. Stir in the capers, lemon zest and juice, lemon slices and stock. Increase the heat and allow the sauce to bubble away for 5 minutes, or until it is slightly reduced and thickened. Return the chicken to the pan and add the parsley. Season to taste.

4. While the chicken is cooking, pulse the cauliflower florets in a food processor until they resemble grains of rice. Brush a frying pan with the remaining olive oil and set over a medium heat. Add the chilli flakes and cauliflower and cook for 3 minutes, stirring well, until heated through. Fluff up with a fork and season with salt and pepper.

5. Divide the chicken and lemon slices in their sauce between 2 serving plates. Serve with the cauliflower rice and rocket.

Lemony Seared Chicken & Veggie Noodles

Spiralised carrots make a colourful and delicious change to courgetti with their slightly sweet flavour. It's fun experimenting with different vegetables to find out which are your favourite veggie noodles. Try beetroot, butternut squash, celeriac, broccoli stems, kohlrabi, parsnips, swede and sweet potato.

SERVES 2
PREP: 15 MINUTES
COOK: 10–12 MINUTES

160g carrots, peeled
2 tbsp olive oil
260g chicken breast fillets, skinned
10g spring onions, thinly sliced
2 garlic cloves, crushed
a pinch of crushed chilli flakes* (optional)
juice of 1 large lemon
100ml good-quality vegetable or chicken stock
50g small broccoli florets
40g cherry tomatoes*, quartered
1 small bunch of parsley, finely chopped
lemon wedges, to garnish
sea salt and freshly ground black pepper

VARIATIONS

- Try using courgettes, swede or butternut squash instead of carrots (squash will take longer to cook).
- Use mushrooms instead of tomatoes.
- Use turkey instead of chicken.

1. Spiralise the carrots lengthways using a spiraliser, julienne peeler, mandolin slicer or potato peeler. Set aside.

2. Heat 1 tablespoon of the oil in a griddle pan set over a medium heat. Put the chicken fillets into the hot pan and cook them for 10–12 minutes, stirring occasionally, until they are golden brown all over and cooked through.

3. Meanwhile, heat the remaining oil in a large frying pan and cook the spring onions and garlic over a low heat, stirring occasionally, for 2–3 minutes until softened. Add the chilli flakes (if using), lemon juice and most of the stock and bring to the boil. Reduce the heat and stir in the broccoli florets and tomatoes and half the parsley. Simmer gently for 2–3 minutes until the sauce reduces and the broccoli is just tender. If the sauce is too thick, thin with more stock.

4. Cut the chicken fillets into slices or chunks and add them to the pan with the carrot strands and the remaining parsley. Cook for 1–2 minutes until the carrots are tender. Season to taste with salt and pepper and divide between 2 serving bowls.

Steak With Sweet Potato Fries

Homemade sweet potato fries taste as good as (if not better than) the real thing. They're healthy, easy to make and you can flavour them with herbs and spices. They taste great with a juicy steak.

SERVES 2
PREP: 15 MINUTES
COOK: 30 MINUTES

180g sweet potato, peeled and cut into thin chips
2 tbsp olive oil
2 x 130g lean sirloin or fillet steaks
40g wild rocket
40g cherry tomatoes*, quartered
sea salt

CAJUN SEASONING
½ tsp black peppercorns
½ tsp white peppercorns
1 tsp sweet paprika*
½ tsp cayenne pepper*
½ tsp garlic powder
1–2 tsp mixed dried herbs, e.g. thyme, oregano, sage

1. Preheat the oven to 200°C/180°C fan/gas mark 6.

2. Make the Cajun seasoning: grind the peppercorns in a pepper grinder or pestle and mortar and mix them with the other ingredients.

3. Place the sweet potato chips in a bowl with 1 tablespoon of the oil and turn them in the oil until they are lightly coated. Sprinkle with the Cajun seasoning and turn again.

4. Spread out the chips in a single layer on a baking tray and add a little sea salt. Cook in the preheated oven for about 30 minutes, turning once or twice, until they are golden brown and crisp outside and tender inside.

5. Meanwhile heat the remaining olive oil in a non-stick griddle pan or frying pan over a medium to high heat. Put the steaks into the hot pan and cook them to your liking: for 2–3 minutes each side (rare); 4–5 minutes (medium); or 5–6 minutes (well done).

6. Serve immediately with the sweet potato fries, rocket and cherry tomatoes.

VARIATIONS

- You can use swede instead of sweet potato.
- Dust the sweet potato fries before baking with ground cumin, coriander and smoked paprika instead of the cajun seasoning.
- Instead of spices, just sprinkle with garlic powder and dried herbs.
- Sprinkle the cooked fries with chopped parsley.

TIP: You can use wooden or metal skewers. If using wooden ones, soak them in water first to prevent them burning.

Beef Kofta Skewers

Spicy koftas make a delicious alternative to beef burgers. You can make them in advance and store them in a sealed container in the fridge until you're ready to cook them later the same day. These are served with salsa but you could try a salad or some cooked green vegetables instead.

SERVES 2
PREP: 15 MINUTES
COOK: 10–12 MINUTES

260g minced beef
60g red onion, grated
1 garlic clove, crushed
1 red chilli*, deseeded and diced
½ tsp ground cumin
½ tsp ground coriander
a handful of coriander, chopped
2 tbsp olive oil
100g courgettes, thickly sliced
sea salt and freshly ground black pepper

SPICY TOMATO SALSA
40g red onion, diced
60g juicy tomatoes*, diced
1 red chilli *, deseeded and diced
a handful of coriander, chopped
juice of 1 small lemon

1. Make the salsa: mix all the ingredients together in a bowl, adding salt to taste.

2. Put the beef, onion, garlic, chilli, ground spices and coriander in a bowl with some salt and pepper. Mix well and then use your hands to mould the mixture into 6 sausage shapes. Thread them onto skewers and brush them lightly with 1 tablespoon of oil.

3. Cook the koftas under a hot grill or in a griddle pan, turning them occasionally, for 6–8 minutes, or until they are browned and cooked through.

4. Heat the remaining oil in a frying pan or griddle pan and cook the courgettes for 2 minutes on each side, or until tender and golden.

5. Serve the koftas and courgettes with the spicy tomato salsa.

VARIATIONS

- Use mint or flat-leaf parsley in the kofta mixture.
- Instead of courgettes, use sliced fennel, red peppers*, mushrooms, butternut squash or pumpkin.
- Serve the koftas with an Indian kachumber salad (see page 111).

NOTE

You can also cook the koftas on a hot barbecue.

Quick Herby Chicken & Vegetable Parcels

When you're rushing and need to cook a speedy meal, this is the perfect solution. Just wrap the chicken and vegetables in paper or foil parcels and pop them into the oven for 30 minutes. Hey presto! Supper is served – it doesn't get any easier than this.

SERVES 2
PREP: 15 MINUTES
COOK: 25–30 MINUTES

260g chicken breast fillets, skinned
1 tsp grated fresh ginger
2 tbsp olive oil
2 tbsp tamari
50g red or yellow pepper*, deseeded and cut into chunks
100g mushrooms, quartered
100g small broccoli florets
10g spring onions, thinly sliced
freshly ground black pepper

1. Preheat the oven to 200°C/180°C fan/gas mark 6.

2. Gently mix the chicken cubes and ginger in a bowl with the olive oil and tamari.

3. Cut out 2 large squares of baking parchment or kitchen foil and divide the vegetables between them. Arrange the chicken cubes on top and drizzle them with any sauce left in the bowl. Season with black pepper and fold the edges of the baking parchment or foil over the top so they meet in the middle. Twist them together to make 2 sealed parcels and place them on a baking tray.

4. Bake in the oven for 25–30 minutes, or until the chicken is cooked right through and the vegetables are tender. Serve immediately.

TIP: You probably won't need to add salt as tamari is quite salty.

VARIATIONS

- Vary the vegetables: use courgette, cauliflower, baby carrots, celery or fennel.
- Use white fish or salmon fillets instead of chicken.
- For a more intense flavour, prepare the chicken in advance and marinate in a sealed container in the fridge until you're ready to assemble and cook the parcels.

Griddled Tofu With Salsa Verde

Use firm tofu for griddling, grilling and frying. The key to achieving a crisp, golden brown exterior and a tender, moist interior when grilling or frying is to press it before using. Place the slices on a baking tray in a single layer between 2 sheets of kitchen paper. Cover with a clean cloth and some heavy cans. Leave for at least 30 minutes, or until any water has drained out.

SERVES 2
PREP: 15 MINUTES
SOAK: 10 MINUTES
COOK: 8–12 MINUTES

100g asparagus, trimmed
40g thin spring onions, trimmed
120g courgette, trimmed and sliced lengthways
juice of ½ lemon
1 tbsp olive oil
260g firm tofu, cut into slices and pressed
sea salt and freshly ground black pepper

SALSA VERDE
1 tbsp dried wakame seaweed
1 garlic clove, crushed
1 tsp capers, rinsed, drained and diced
a few sprigs of flat-leaf parsley, chopped
a few sprigs of coriander, chopped
grated zest and juice of ½ unwaxed lemon
1 tbsp olive oil

VARIATIONS

- Enliven the salsa verde with diced chilli*, crushed chilli flakes* or a chopped anchovy fillet.
- Vary the griddled vegetables: try squash, aubergine*, fennel and purple-sprouting broccoli.

1. Make the salsa verde: put the wakame seaweed in a small bowl and cover it with boiling water. Soak it for 10 minutes and then drain. Place it in a bowl with all the other ingredients and mix well. Season with salt and pepper and set aside. If it looks too thick you can thin it with a little water.

2. Place the asparagus, spring onions and courgette pieces in a bowl with the lemon juice, olive oil and some salt and pepper. Toss the vegetable pieces very gently until they are lightly coated.

3. Set a ridged non-stick griddle pan over a medium heat and when it's hot, add the vegetables (cook in batches if the pan is not very large). Cook them for 2–3 minutes on each side, or until they are just tender and starting to char. Remove from the pan and keep warm.

4. Cook the sliced tofu in the hot griddle pan for 2–3 minutes, or until golden. Carefully turn them over and cook the other side. Remove and drain on kitchen paper.

5. Divide the griddled spring vegetables between 2 serving plates and top with the tofu. Drizzle with the salsa verde and serve immediately.

Garlicky Prawn Courgetti

This is my take on a traditional Sicilian dish. You don't need to use fresh prawns – the packs of frozen tiger prawns in the freezer aisle of your local supermarket will taste delicious. Be sure to defrost them thoroughly before cooking.

SERVES 2
PREP: 10 MINUTES
COOK: 10–12 MINUTES

200g courgette, ends trimmed
2 tbsp olive oil
3 garlic cloves, crushed
a pinch of dried chilli flakes*
a handful of flat-leaf parsley, finely chopped
a few chives, snipped
juice of 1 large lemon
120ml good-quality vegetable stock
60g cherry tomatoes*, diced
260g peeled raw tiger prawns
sea salt and freshly ground black pepper

1. Use blade C of a spiraliser to cut the courgette lengthways into long thin strands. If you don't have a spiraliser, slice it very thinly lengthways with a julienne peeler, mandolin slicer or a potato peeler.

2. Heat the oil in a large frying pan and set over a medium heat. Add the garlic and cook for 1 minute without allowing it to colour. Stir in the chilli flakes and half the chopped herbs, then pour in the lemon juice and stock and bring to the boil. Let the mixture bubble away for 4–5 minutes until the liquid reduces.

3. Add the tomatoes and prawns and cook, turning once, for 2–3 minutes, or until the prawns are pink on both sides. Gently stir in the courgette strands and cook for 1–2 minutes. Season to taste with salt and pepper and stir in the remaining herbs.

4. Divide between 2 serving plates or shallow bowls and serve immediately.

VARIATIONS

- Use scallops instead of prawns.
- Vary the herbs: try chopped coriander, basil or dill.
- Instead of adding tomatoes, serve with steamed or boiled broccoli or thin green beans.
- Strew with fresh rocket leaves or stir some in just before serving.

PHASE 4
Forever

Phase 4: Forever

- **How does your new life look?**
- **What do you need to keep doing to keep yourself looking and feeling as good as you now do?**

HBDers often tell me they miss the safety and security of the Phase 3 rules and that entering Phase 4 makes them feel all at sea. So think of Phase 4 as an extension of Phase 3. You're building on the feedback your body has given you and discovered, by reintroducing certain foods in Phase 3 and via your treat meals, which foods best suit you and which are best left alone.

This is the time to experiment with more carbohydrates if you'd like to (especially gluten-free grains such as rice, buckwheat and quinoa) and potatoes from time to time, and essentially to see what you can get away with. You can have some extra treat meals, too. Cottage cheese, chia seeds and all the other 'diet foods' can be reintroduced if you're so inclined (but not miserable egg white omelettes, of course!).

Feel free to experiment with mixing proteins sometimes, mixing fruits sometimes, or adding in other foods that have been excluded, for example different grains. Personally I avoid grains because I feel so much better without them. So build on the lessons you've learned over the past weeks about which foods best suit you as an individual. You may well find that a tablespoon of rice, quinoa or buckwheat suits you just fine. Continue to experiment but listen to the feedback and messages from your body. Some experienced Phase 4 HBDers still keep their food diary going because it's so helpful to keep tabs on what best suits us and to avoid getting bogged down by eating in a repetitive way.

Reintroduce offal/liver, if you'd like to, and occasional sourdough bread if it suits you. Why sourdough? Because the fermentation process means there's much less gluten in it. At Riccardo's (my husband's restaurant in Chelsea) the chef, Paolo, started experimenting with making his own sourdough bread a couple of years ago and several staff members (who found themselves bloated with normal bread) discovered that they could happily eat Paulo's delicious new sourdough bread.

In practice, when eating at home most HBDers adhere to the one protein per meal rule – so no salad niçoise or smoked salmon and scrambled eggs – and enjoy occasional sourdough bread, or a glass of wine when out with friends for lunch. Black coffee or black or green tea occasionally between meals is fine, too.

Experiment with adding in extra fruit, things like oranges, grapefruit, banana and kiwi, but always bear in mind that it's the vegetables in our daily diet that are important, not the fruit. Personally, other than an apple, I rarely eat fruit – it makes me feel hungry and my body doesn't appreciate it.

The number one Phase 4 rule, and pretty much the only rule (other than to keep drinking the water and to keep sugar to a minimum and stay off the foods that you discovered don't suit you in Phase 3), is to never ever to go back to snacking in between meals.

Experiment with skipping a meal now and then, especially breakfast, also experiment with eating one meal a day (OMAD); it's tough but strangely satisfying. Some people find occasional fasting really suits them – this is the phase for that. It's a good thing to shake things up and to take our body by surprise now and again. You no longer need to start your meal with protein because our blood sugar levels have now stabilised – have salad or vegetables first if you'd like to.

These are the rules that are with us forever, the rules that we follow to protect both our health and our waistline:

1. Always start the day with water: 500ml and drink another 1.5 litres before lunch.

2. Always leave at least 5 hours between finishing one meal and starting the next.

3. Always finish eating within an hour (unless it's a treat meal) in order to avoid falling back into the grazing/snacking trap.

4. Always eat your apple a day, with a meal.

This is the time to reintroduce other foods back into your diet but continue to listen to feedback from your body. Keep your food/symptom diary going. Remember your meals are primarily still focused on protein and vegetables – keep up all the good habits you have made in Phases 2 and 3. Avoid introducing too much in the way or grains/potato/sweetcorn into your daily diet.

PHASE 4 CORE ACTIONS

In general, have one type of protein per meal – but occasional protein mixing is fine.

Quantities are more relaxed in Phase 4 – but we're still on a mission to find out what suits us best and makes us feel good!

All permitted foods as per Phases 1–3, with additions below

Reintroducing some natural/unprocessed foods
• Fats and oils: butter, goose/duck fat, avocado, sesame and coconut oil (but remember that EVOO reigns supreme)
• All nuts and seeds (peanuts are legumes rather than nuts – see how you feel on them) and nut butters
• All fruit and occasional mixes of fruit
• Honey: small amounts now and again are fine
• Plant yoghurts and dairy substitutes e.g. coconut yoghurt, milk and cream
• Potatoes
• Sweetcorn
• Peas
• 80% cocoa content dark chocolate – 2 squares a day after a meal

Banned Phase 4 Foods

- No vegetable oils, including rapeseed oil
- No processed foods – remember to read labels and avoid nutritionally dead 'factory food'
- No more than a tiny amount of sugar – don't let it sneak back in and become a habit. Remember it's an 'anti-nutrient'

EXPERIMENT WITH

Grains – discover what suits you. You could try buckwheat, quinoa, rice oats (including porridge oats for breakfast), amaranth and anything else that takes your fancy and see how your body reacts to them. The key is to include small amounts to accompany your meal of protein and veg rather than to base your meal on these nutritionally poor carbs.

Wheat – some of us cope well with occasional gluten-grains, including spelt and kamut, but others have to steer clear. Listen to your body.

PHASE 4 BREAKFASTS

No-Pastry Mini Quiches

These little quiches loaded with colourful vegetables make a very healthy and satisfying breakfast. They're gluten-free and practically carb-free too as there's no pastry.

SERVES 2
PREP: 15 MINUTES
COOK: 30 MINUTES

2 tbsp olive oil
60g red onion, diced
50g red or yellow pepper*, deseeded and diced
50g button mushrooms, diced
100g baby spinach leaves
4 organic eggs
4 tbsp milk
a handful of flat-leaf parsley, chopped
60g feta cheese, crumbled
sea salt and freshly ground black pepper

1. Preheat the oven to 190°C/170°C fan/gas mark 5.

2. Heat most of the olive oil in a frying pan over a medium heat. Add the onion and pepper and cook for 6–8 minutes until they soften. Stir in the mushrooms and cook for 2–3 minutes. Add the spinach and cook for 1 minute, or until it wilts and turns bright green.

3. Lightly brush 4 non-stick muffin pans with the remaining oil and divide the vegetable mixture between them.

4. Beat the eggs and milk in a bowl. Stir in the parsley and feta cheese and season with salt and pepper.

5. Pour the egg mixture over the vegetables, then bake the mini quiches in the oven for 20 minutes, or until they are risen, golden brown and firm to the touch.

6. Run the blade of a thin knife around the inside of the muffin pans to unmould the quiches. Eat them warm or set them aside to cool and then store in a sealed container for breakfast the following day.

VARIATIONS

- Use grated Cheddar instead of feta cheese.
- Vary the vegetables: try spring onions, shallots, tomatoes*, courgette, kale or leeks.
- Instead of parsley, try chives, thyme or oregano.

TIP: You could enjoy these quiches for a delicious packed lunch.

Chia Seed Pudding

Chia seeds are not on my personal list of favourite foods but they're certainly popular. Anything that miraculously absorbs that much water and transforms itself into a gelatinous mass just doesn't appeal. But if you've reached Phase 4 and you've been longing for chia seed pudding for breakfast, this one's for you – enjoy!

SERVES 2
PREP: 10 MINUTES
CHILL: OVERNIGHT

4 tbsp chia seeds
240ml dairy or non-dairy milk, e.g. unsweetened almond or coconut (depending on what suits you)
2 tbsp unsweetened dairy or non-dairy yoghurt, e.g. soya or coconut
clear honey or maple syrup, for drizzling (optional)
berries for topping, e.g. strawberries, raspberries, blueberries, blackberries

1. Take 2 glass jars with screwtop lids and put 2 tablespoons chia seeds and 120ml milk into each one. Mix together or whisk well until everything is combined. Leave for 3–4 minutes and then stir or whisk again to ensure that the chia seeds are well distributed throughout and not sticking together.

2. Screw the lids on the jars and chill them in the fridge overnight.

3. The following morning, unscrew the lids, place a spoonful of yoghurt on top of each pudding and drizzle them with honey (if using). Top with fresh berries and enjoy!

TIP: These chia seed puddings will keep well in the fridge for up to 5 days.

VARIATIONS

- Top with sliced or diced banana, peach, nectarine, papaya, kiwi or mango.
- Mix the honey into the chia seed mixture before chilling.
- Sprinkle with coconut flakes or chopped nuts.
- Sprinkle with fresh pomegranate seeds.
- Flavour the puddings with grated orange zest.
- Whisk a small mashed banana into the chia seeds and milk before chilling.

PHASE 4
MAINS

TIP: Instead of using a grill, you can cook the cod in a preheated oven at 200°C/180°C fan/gas mark 6. The cooking time will be the same.

Japanese Black Cod

If you love Japanese food, this is a great dish to make at home. You need to plan it in advance as the cod is marinated overnight. It's quick and easy to make and perfect for entertaining, too.

SERVES 2
PREP: 15 MINUTES
MARINATE: OVERNIGHT
COOK: 15 MINUTES

2 x 130–175g thick cod fillets, skinned
2 tbsp olive oil
150g broccoli florets
150g sugar snap peas
2 tbsp teriyaki sauce
1 tsp white sesame seeds
2 spring onions, shredded
2 tsp pickled ginger, to serve

MARINADE
2 tbsp sake
2 tbsp mirin
3 tsp clear honey
2 tbsp white miso paste

1. Make the marinade: put the sake, mirin and honey in a small pan over a high heat. Stir gently and when the mixture comes to the boil, reduce the heat to low and add the miso paste. Stir until it dissolves and then remove the pan from the heat. Set aside to cool.

2. Pour the cooled marinade into a container and add the cod fillets. Turn them in the marinade until they are coated. Cover with a lid or cling film and marinate in the fridge overnight.

3. When you're ready to cook the cod, pat it dry with kitchen paper. Heat the oil in a large frying pan over a medium to high heat. When the pan is hot, add the cod fillets and cook them for 2–3 minutes, or until they brown underneath. Turn them over gently and cook for 2–3 minutes on the other side.

4. Place the pan under a hot grill and cook for 5–10 minutes, or until the cod is flaky and cooked through.

5. Meanwhile, boil, steam or microwave the broccoli and sugar snap peas until they are just tender. Toss them in the teriyaki sauce and sprinkle with sesame seeds.

6. Serve the cod topped with shredded spring onions, with the broccoli and sugar snap peas, and some pickled ginger on the side.

VARIATIONS

- Vary the vegetables: try fine green beans, pak choi or asparagus.
- Use tamari sauce instead of teriyaki.

Roasted Sweet Potato Salad Bowl

This versatile roasted vegetable and crunchy bulgur wheat salad can be enjoyed lukewarm or cold. Bulgur wheat has a pleasantly nutty texture, but if wheat doesn't suit you, or you are avoiding gluten, use quinoa or buckwheat instead. You can also substitute pumpkin or butternut squash for the sweet potatoes. Keep the salad overnight in a sealed container in the fridge to take to work as a packed lunch.

SERVES 2
PREP: 15 MINUTES
COOK: 30–35 MINUTES

300g sweet potatoes, peeled and cubed
1 red onion, cut into thin wedges
2 tbsp olive oil
100g bulgur wheat (dry weight)
125ml good-quality vegetable stock
juice of 1 lemon
a handful of flat-leaf parsley or mint, chopped
2 tbsp toasted pine nuts
1 tbsp pumpkin seeds
a small bunch of rocket
100g creamy goat's cheese, crumbled
sea salt and freshly ground black pepper

TAHINI DRESSING
2 tbsp tahini
2 tbsp yoghurt
1 garlic clove, crushed
juice of ½ lemon

1. Preheat the oven to 200°C/180°C fan/gas mark 6.

2. Toss the sweet potato cubes and onion wedges in 1 tablespoon of the oil, then spread them out on a baking tray. Season lightly with salt and pepper. Roast for 30–35 minutes until they are golden brown.

3. Meanwhile, put the bulgur wheat and stock in a saucepan and bring it to the boil. Reduce the heat, cover the pan and simmer for 5 minutes. Remove from the heat and leave in the covered pan for at least 5 minutes until the bulgur wheat is tender and has absorbed most of the stock. Stir in the lemon juice, remaining olive oil and chopped herbs.

4. Meanwhile, make the tahini dressing: mix the tahini, yoghurt, garlic and lemon juice in a bowl. It should have the consistency of pouring cream. If it's too thick add a little water. Season to taste with salt and pepper.

5. Set a dry frying pan over a medium heat and toast the pine nuts and pumpkin seeds for 1–2 minutes, tossing them gently. Remove before they burn.

6. Divide the bulgur wheat between 2 shallow bowls and top with the roasted sweet potato and red onions. Add the rocket and drizzle with the tahini dressing. Crumble over the goat's cheese and sprinkle with the toasted pine nuts and seeds.

Quick Mozzarella Mushroom Chicken

Now you're on Phase 4 you can mix your proteins occasionally when you're cooking at home. This is one of the easiest suppers ever – apart from the vegetable accompaniment, everything is cooked in one pan, which makes washing up much quicker!

SERVES 2
PREP: 5 MINUTES
COOK: 15–20 MINUTES

2 tbsp olive oil
2 x 130g chicken breast fillets, skinned
200g chestnut mushrooms, sliced
2 x 30g slices of mozzarella
1 tsp diced sun-dried tomatoes*, drained (optional)
80g fine green beans or purple sprouting broccoli, steamed
snipped chives or sliced spring onions, to garnish
sea salt and freshly ground black pepper

1. Heat the oil in a large frying pan over a medium heat. When it's hot, add the chicken breasts and cook them for 2–3 minutes on each side until they are golden brown.

2. Add the mushrooms and cook them gently, stirring occasionally, for 5 minutes or until they are tender and golden. Season with salt and pepper.

3. Put a slice of mozzarella on top of each chicken breast and add the sun-dried tomatoes (if using). Cover the pan with a lid or a sheet of kitchen foil and cook for 5 minutes, or until the mozzarella melts and the chicken is cooked through. Season lightly with salt and pepper.

4. Serve immediately, sprinkled with chives or spring onions, with some steamed green beans or broccoli.

TIP: The chicken breasts will cook more quickly if you flatten them first with a meat mallet or rolling pin.

VARIATIONS

- Add a slice of tomato* and a sprig of fresh basil with the mozzarella before covering the pan for the last 5 minutes.
- Drizzle the cooked chicken with pesto or oil-free pesto (see page 78).
- Serve with a crisp salad instead of cooked green vegetables.

Green Quinoa Tabbouleh

Tabbouleh is traditionally made with bulghur wheat but it also works well with quinoa, which is more nutritious and gluten free. Cooking it in vegetable stock instead of water gives it a delicious flavour. Pistachio, pomegranate seeds and pine nuts add nutritional value as well as fabulous textures.

SERVES 2
PREP: 20 MINUTES
COOK: 15 MINUTES

75g quinoa (dry weight)
200ml good-quality vegetable stock
30g shelled pistachios
50g curly kale, roughly chopped
50g baby spinach leaves, shredded
30g wild rocket, chopped
4 spring onions, chopped
a handful of coriander, chopped
1 small avocado, peeled, stoned and diced
1 tbsp toasted pine nuts
½ tsp ground allspice
seeds of ½ pomegranate

DRESSING
1 garlic clove, crushed
1 tsp white sesame seeds
1 small green bird's eye chilli*, shredded or diced (optional)
a pinch of sea salt
2 tbsp fruity green olive oil
juice of 1 small lemon

1. Rinse the quinoa in a sieve under cold running water. Bring the vegetable stock to the boil in a saucepan and stir in the quinoa. Reduce the heat, cover the pan and simmer for 12–15 minutes until just tender. Turn off the heat and leave the quinoa to steam in the pan for 5 minutes. Drain well, fluff up with a fork and set aside to cool.

2. Set a non-stick frying pan over a medium heat. Put the pistachios into the hot pan and toast them, tossing gently, for 1–2 minutes until they are golden. Watch them to make sure they don't burn. Remove from the pan and cool before chopping.

3. Make the dressing: in a pestle and mortar crush the garlic, sesame seeds, chilli and salt. Gradually stir in the oil to make a garlicky paste, then add the lemon juice, stirring until well amalgamated.

4. Mix together the kale, spinach, rocket, spring onions and coriander in a bowl. Gently stir in the avocado, pine nuts, allspice, cooled pistachios and quinoa. Toss everything together in the dressing.

5. Divide the tabbouleh between 2 serving bowls. Sprinkle with pomegranate seeds and serve.

VARIATIONS

- Top with grilled halloumi or tofu.
- Add roasted red or yellow peppers*, butternut squash or pumpkin.

Stuffed Mediterranean Vegetables

Baked vegetables filled with other vegetables and rice or grains are popular in southern Italy, Greece and the Levant. If you don't like aubergines or peppers, you can use hollowed out courgettes or large beefsteak tomatoes* instead.

SERVES 2
PREP: 15 MINUTES
COOK: 30 MINUTES

1 large aubergine*
1 large red or yellow pepper*
2 tbsp olive oil, plus extra for drizzling
75g buckwheat (dry weight)
1 small red onion, diced
2 garlic cloves, crushed
2 tomatoes*, diced
a handful of flat-leaf parsley, chopped
2 tbsp toasted pine nuts
grated zest and juice of 1 small lemon
100g feta cheese, diced
sea salt and freshly ground black pepper

VARIATIONS

- Stir some sunflower or pumpkin seeds into the buckwheat for more crunch.
- Add a spoonful of caraway seeds.
- If you don't have feta, sprinkle the stuffed vegetables with grated Parmesan or Cheddar before popping them back into the oven for the final 10 minutes.

1. Preheat the oven to 200°C/180°C fan/gas mark 6.

2. Cut the aubergine and peppers in half top to bottom through the stalk. Discard the white ribs and seeds from the peppers. Place the aubergines and peppers cut-side up on a baking sheet and drizzle them with 2 teaspoons of the oil. Bake in the oven for 20 minutes, or until they are tender. Remove and cool a little – but don't turn off the oven.

3. Meanwhile, cook the buckwheat according to the instructions on the packet.

4. While it's cooking, heat the remaining oil in a frying pan and cook the red onion and garlic over a low heat, stirring occasionally, for 10–15 minutes, or until they are softened but not coloured. Stir in the tomatoes and parsley and cook gently for 4–5 minutes. Season with salt and pepper.

5. Scoop out and dice the cooked aubergine flesh, putting the skin to one side. Mix the diced flesh into the onion and tomato mixture with the cooked buckwheat, pine nuts, lemon zest and juice and feta, and stir well. Season to taste with salt and pepper.

6. Pile the mixture into the aubergine and pepper shells and return them to the hot oven for 8–10 minutes until they are golden brown. Serve warm or cold with salad.

TIP: To toast pine nuts, spread them out in a dry frying pan over a low heat and cook for 1–2 minutes, shaking gently, until they are golden brown and fragrant. Remove before they burn.

Miso Broth

This umami-tasting soup is healthy, cleansing and full of nutritional goodness, and it keeps well in a sealed container in the fridge for a couple of days. The flavour is only as good as the stock you use, so make sure it's the best possible, preferably homemade. To make it even more nutritious use chicken rather than vegetable stock.

SERVES 4
PREP: 10 MINUTES
SOAK: 10–15 MINUTES
COOK: 15 MINUTES

1 tbsp dried wakame seaweed
1 litre good-quality vegetable stock
4 tbsp white miso paste
6 spring onions, thinly sliced
2 garlic cloves, thinly sliced
1 tsp grated fresh ginger
2 celery sticks, chopped
250g sliced shiitake mushrooms
100g thin asparagus stems, trimmed and sliced
1 pak choi, trimmed and leaves separated
200g silken tofu, cut into cubes
a dash of lime juice (optional)
tamari sauce, to taste (optional)
a small handful of coriander, mint or basil, chopped

1. Put the wakame in a bowl and cover it with plenty of cold water. Leave for 10–15 minutes until it is soft and rehydrated. Drain well.

2. Meanwhile, pour the stock into a large saucepan and bring to the boil. Reduce the heat to low and add the miso paste. Stir with a wooden spoon until it dissolves.

3. Add the spring onions, garlic, ginger, celery, mushrooms and asparagus. Simmer gently for 4–5 minutes, then stir in the pak choi and cook for 2–3 minutes. Add the wakame and tofu and heat through gently for 1 minute. Check the seasoning, adding lime juice and tamari sauce to taste, if wished.

4. Divide the broth between 4 shallow bowls, sprinkle with herbs and serve immediately.

VARIATIONS

- For a hot and sour flavour add 1 teaspoon of unsweetened rice vinegar at the end.
- Add some shredded spinach, Brussel sprout tops or spring greens.
- Add some thin rice noodles.
- Instead of tofu, use cooked chicken breasts, cut into chunks.

A

B

C

D

E

F

G

H

M

N

O

P

Q

R

T

U

V

W

Y

Z

Huge gratitude to my editor at HarperCollins, Katya Shipster, and to her wonderful team, who have made the *HBD Cookbook* a reality. And more than that, a beautiful reality thanks to the talented Catherine Wood. Thanks to Katya and her team, your pleas, dear HBDers, have been answered – a real cookbook at last, to make your HBD experience easier and more delicious.

And heartfelt thanks to the HBD Gang on Instagram for spreading the love and inspiration; without you this book would never have been born. And a special thank you to my dear friend, Lara Ross.